D0945821

2

About Island Press

Since 1984, the nonprofit Island Press has been stimulating, shaping, and communicating the ideas that are essential for solving environmental problems worldwide. With more than 800 titles in print and some 40 new releases each year, we are the nation's leading publisher on environmental issues. We identify innovative thinkers and emerging trends in the environmental field. We work with world-renowned experts and authors to develop cross-disciplinary solutions to environmental challenges.

Island Press designs and implements coordinated book publication campaigns in order to communicate our critical messages in print, in person, and online using the latest technologies, programs, and the media. Our goal: to reach targeted audiences—scientists, policymakers, environmental advocates, the media, and concerned citizens—who can and will take action to protect the plants and animals that enrich our world, the ecosystems we need to survive, the water we drink, and the air we breathe.

Island Press gratefully acknowledges the support of its work by the Agua Fund, Inc., The Margaret A. Cargill Foundation, Betsy and Jesse Fink Foundation, The William and Flora Hewlett Foundation, The Kresge Foundation, The Forrest and Frances Lattner Foundation, The Andrew W. Mellon Foundation, The Curtis and Edith Munson Foundation, The Overbrook Foundation, The David and Lucile Packard Foundation, The Summit Foundation, Trust for Architectural Easements, The Winslow Foundation, and other generous donors.

The opinions expressed in this book are those of the author(s) and do not necessarily reflect the views of our donors.

Resilience Practice

LIBRARY OF ROWAN COLLEGE
AT BURLINGTON COUNTY

Resilience Practice

Building Capacity to Absorb Disturbance and Maintain Function

Brian Walker & David Salt

WASHINGTON | COVELO | LONDON

Copyright © 2012 Brian Walker and David Salt

All rights reserved under International and Pan-American Copyright Conventions.
No part of this book may be reproduced in any form or by any means
without permission in writing from the publisher:
Island Press, 1718 Connecticut Avenue NW, Suite 300, Washington, DC 20009.

ISLAND PRESS is a trademark of The Center for Resource Economics.

Library of Congress Cataloging-in-Publication Data

Walker, B. H. (Brian Harrison), 1940-
  Resilience practice : building capacity to absorb disturbance and maintain
function / Brian Walker, David Salt.
    p. cm.
  ISBN 978-1-59726-800-4 (hardback) -- ISBN 1-59726-800-3 (cloth) -- ISBN
978-1-59726-801-1 (paper) 1. Natural resources--Management. 2. Natural
resources--Management--Case studies. 3. Resilience (Ecology)--Case studies.
4. Nature conservation--Case studies. 5. Environmental protection--Case
studies. I. Salt, David (David Andrew) II. Title.
  HC59.15.W348 2012
  333.7--dc23

                                        2012016122

Printed on recycled, acid-free paper

Manufactured in the United States of America
10 9 8 7 6 5 4 3 2 1

Keywords: Island Press, complex adaptive systems, sustainability, tipping
points, planetary boundaries, adaptive management, adaptive governance
and state-and-transition models, natural resources management, sustainable
development, human ecology, ecological resilience, resilience thinking,
managing resilience, practicing resilience, resilience assessments, specified
resilience, general resilience, transformability, self-organizing systems,
adaptive cycles, adaptive systems, ecological threshold, panarchy, ecosystem
services, resilience.

Contents

Foreword

This is a timely and compelling book. The future of our planet and of ourselves is looking increasingly uncertain. We are beset by stresses and shocks—of all kinds, natural and human induced—that are growing in frequency and size. We have shown enormous ingenuity in the past in applying science and technology to increase food production, reduce mortality, and improve the quality of human life, even though the benefits of these improvements have not always been shared equally around the planet. But we've been less effective in managing our impacts on the environment, whether in our backyard or for the planet as a whole.

This book is in some respects a sequel to Brian Walker and David Salt's 2006 book *Resilience Thinking*. Since the publication of that book, the number of serious environmental events and unwanted changes occurring in ecosystems, farming regions, forests, and the oceans has increased, as the world approaches planetary boundaries. And as people have begun to understand the severity of the challenges we face, there is growing interest in the concept of resilience, with more and more people wondering what might happen, and whether we could cope, if and when some of the looming shocks strike us.

Resilience thinking has emerged as a valuable way for people to engage with the world. Indeed, interest has reached the point where the term resilience is considered by some to be the "new sustainability" and is developing into a buzzword. Its increasingly common use in political rhetoric involves various interpretations of what it means and carries the danger of its value being discounted.

This book is a practical primer. It takes the reader through the basics that underpin resilience thinking and then sets out how this valuable set of ideas might actually be applied in assessing and managing resilience. Chapters on how an assessment might be approached are interspersed with case studies that describe how resilience applies in a range of real-world situations.

Underlying resilience, in theory and practice, is the need to see the world as consisting of a large number of different systems—small and large, natural and physical, often combined in complex ways—on which we depend. It focuses on the changes, and consequences of management actions, that matter most in these systems. To be effective we need to understand how our activities in one part of the system affect other parts, better engage stakeholders to play active roles in working with their systems, and assist in designing fair and robust structures of governance that facilitate that engagement. Big challenges indeed, but we've demonstrated the power of systems thinking in the past.

The Green Revolution that brought about dramatic increases in food production, increases that were able to keep up with population growth, was an example of a systems approach in practice. New genes were bred into wheat and rice varieties that made them able to take up high dosages of fertilizer and produce high yields. Alongside, the logistics of supplying large quantities of fertilizers, pesticides, and water were put into place. Farmers in the developing countries responded eagerly, and yields grew dramatically. But some of the key linkages in these systems were ignored. For example, the pesticides, while only partially effective at killing the pests, were very effective at killing their enemies, various parasites and predators. As a result the pests exploded, causing billions of dollars of damage. The Green Revolution was poorly resilient to the effects of modern pesticides, and the system only recovered when integrated pest management practices were adopted.

The Green Revolution is but one example. Others, discussed in this book, include managing livestock grazing and wetlands, designing and managing irrigation systems, and overseeing marine fisheries.

Understanding such system dynamics and the role of resilience enables managers to better deal with these problems. As a result, resilience thinking is now emerging as a valuable process for engaging with the complexity of the systems around us. This book makes a valuable contribution to our efforts to prepare for the growing challenges confronting our planet as the twenty-first century unfolds. Our current trajectory is imperiling future generations, our children and their children, as we approach and overshoot our planet's safe operating limits. We need the thoughts and tools this book provides if we are to avoid catastrophe.

Sir Gordon Conway
Professor of International Development
Imperial College, London

Preface

*"We live in a complex world. Anyone with a stake
in managing some aspect of that world will benefit from a
richer understanding of resilience and its implications."*

These words introduced our earlier book, *Resilience Thinking*. Published in 2006, it formed part of a rising wave of interest in resilience. Part of that interest has arisen from the growing body of research on resilience and natural resource management; part of it stems from a growing disenchantment with more traditional approaches of resource optimization and efforts to maintain "business as usual." There's also an increasing awareness of the consequences of declining resilience in fields like health, economics, the law, and engineering. On top of this there is an increasing worry about society's fragility in the face of a growing catalog of climate catastrophes and natural disasters.

Consider for a moment a few of the events that have shaped our world just recently. Monster tornadoes shredded towns in the United States. Following one of the worst droughts on record, much of eastern Australia was inundated by intense flooding rain. Japan was knocked over by historic earthquakes and then washed away in the subsequent tsunamis. And Europe and North America are recovering from two of the most savage winter storm seasons on record. The world is attempting to understand the consequences of climate change, peak oil, the increasingly volatile financial system, and accelerating declines in biodiversity. The future looks increasingly uncertain.

As the systems that sustain us are subjected to shock after shock, the question that inevitably arises is, How much can they take and still deliver the things we want from them? That, in a nutshell, is the central question behind resilience thinking. In our first book we sought to explain what the science is about and how it adds value to the way we manage the systems around us. In this sequel we continue along that thread and discuss different ways the thinking can be applied.

We also discuss the origins and ideas behind the term itself and the different approaches to how it is used. The word *resilience* is increasingly seen in the lexicons of politicians and leaders. Indeed, it's not uncommon to hear rhetoric such as "we are building resilient communities" or "striving for resilient landscapes."

That fact that resilience is seen as important and is being actively promoted is both a good and a worrying thing. It's good because an honest engagement with the concept of resilience increases our understanding of the systems we are working with.

But raising the profile of an emerging science to the status of a buzzword can also be frustrating as everyone attempts to use the word for their own interests. Buzzwords have a tendency to be all things to all people, and they frequently end up being of little value to anyone. It's been suggested that the term *sustainable* has gone down this road. Some are saying the same thing about the word *resilience*.

Based on the feedback we got from readers of *Resilience Thinking*, there's a large body of support for a resilience approach to managing our landscapes, seascapes, farms, and natural systems. But what's the next step? What do you do with resilience thinking?

The next step is to apply that thinking, to put it into practice. And that's the theme of this book. In its simplest form, it's as basic as ABC, where A has you describing the system, B involves assessing its resilience, and C is about managing that resilience. Of course, as with the best and worst things in life, the devil is in the detail.

In this book we've endeavored to provide an easy-to-read but scientifically robust guide through the world of resilience practice. We've done our best to reduce the jargon (and we've included a glossary to help with some of what we have used), and we've illustrated our discussion with case studies to demonstrate how the lessons might be applicable.

This book is not a second edition of *Resilience Thinking*. It is a companion and sequel. We set the ground in *Resilience Thinking*, and if you haven't read it, you'll get a quick rundown of its essence in the introductory chapter in this book.

However, where we only touched on concepts of specified resilience, general resilience, adaptability, and transformability in *Resilience Thinking*, we spend considerable space here exploring these ideas because understanding what they are and how they are approached is central to any resilience practice.

In *Resilience Thinking* we made the claim that anyone can do it. You don't need a degree or to have spent half your life learning about complex adaptive systems. In *Resilience Practice* we continue in this belief. In your attempt to put resilience thinking into practice, there are many things we can help you with in terms of approaches and frameworks for discussion, but possibly the most important asset you have when it comes to resilience thinking is your own life experience in dealing with the systems around you, and of which you are a part.

Acknowledgments

Many people have assisted with the creation of this book through the contribution of ideas, enthusiasm, and feedback. Their input has considerably improved our output and we deeply appreciate their effort.

Up front we would like to acknowledge Paul Ryan, whose work facilitating a broad range of resilience workshops across Australia has greatly advanced the understanding of how resilience can be operationalized with regional groups. The work he has been doing has informed much of the discussion in chapters 2–4. Having said that, we note that what we present in this book doesn't begin to capture the cultural shift that this work is producing.

Then there's a long list of people who have given their time in providing information, strengthening our arguments, correcting our mistakes, and adding value in a variety of ways to what we've presented. They include Nick Abel, Astier Almedom, Xavier Basurto, William (Buz) Brock, Tim Buchman, Tony Capon, Eddy Carmack, Steve Carpenter, Steve Cork, Michael Cox, Sam Davis, Lisa Deutsch, John Doyle, Andrew Edgar, Carl Folke, Phil Gibbons, Lance Gunderson, C. S. (Buzz) Holling, Miriam Huitric, Steve Lansing, Simon Levin, Rachel and Stephan Lorenzen, Jim McDonald, Per Olssen, Garry Peterson, Allyson Quinlan, Carolyn Raine, Marten Scheffer, Carl Walters, and Frances Westley.

We'd also like to thank our publisher Island Press for helping us polish the manuscript and make it more readable.

And finally, we'd like to acknowledge the steadfast patience and support of our wives, Laura Walker and Yvette Salt. Without that support this book would never have been completed.

1

Preparing for Practice:

The Essence of Resilience Thinking

There are any number of ways of putting resilience science into practice, and it needs to be said at the outset that following strict recipes and prescriptions simply isn't appropriate. Working with resilience requires you to constantly reflect on what you're doing and why you're doing it. And once an assessment of resilience is done, you are encouraged to go back and reexamine it, expand on it, and then adapt accordingly. Our focus in most of this book is on the resilience of social-ecological systems (linked systems of humans and nature). Resilience is a dynamic property of such a system, and managing for it requires a dynamic and adaptive approach.

This being said, the activities undertaken as part of resilience practice can be grouped into three broad steps: *describing* the system, *assessing* its resilience, and *managing* its resilience. In this book we'll provide a variety of ways you can undertake these steps, but the ultimate aim is that you devise your own approach.

While resilience science is not new, attempts to apply it in real-world situations have only recently started taking shape. Workshops of all sizes and flavors have been held around the world on various aspects of resilience practice, and one clear lesson is emerging from this experience. People seeking to undertake resilience assessments or work with resilience need to be in a "resilience frame of mind" to begin with. In other words, it's unlikely they'll engage with resilience practice if they haven't some idea of what resilience is about.

That's not a major hurdle. People with a bit of life experience and some responsibility for managing a system (e.g., a farm, a catchment, a business, or a national park) are usually very quick in picking up on resilience thinking. These systems are self-organizing systems, and people working with them have been attempting to understand them in their day-to-day work. Resilience thinking provides a useful framework for a deeper engagement on why these systems behave as they do.

A simple overview of resilience science is provided in our earlier book, *Resilience Thinking*, but there are also many other resources available at the website of the Resilience Alliance (www.resalliance. org). This is a group of organizations and individuals involved in interlinked aspects of ecological, social, and economic research. It is the network that has created and developed the framework of "resilience thinking."

Resilience and Identity

The word *resilience* is now common in many vision and mission statements. But ask the people who use these statements what they think it means, and you get a range of different answers, most of which relate to how something or someone copes with a shock or a disturbance.

Concepts of resilience are used in all sorts of disciplines, but the term has four main origins—psychosocial, ecological, disaster relief (and military), and engineering. We discuss these in chapter 5, but it's helpful to consider them briefly in this introduction.

Psychologists have long recognized marked differences in the resilience of individuals confronted with traumatic and disastrous circumstances. Considerable research has gone into trying to understand how individuals and societies can gain and lose resilience.

Ecologists have tended to describe resilience in two ways: one focused on the speed of return following a disturbance, the other focused on whether or not the system *can* recover.

People engaging with resilience from the perspective of disaster relief or in a military arena incorporate both aspects (i.e., speed and ability to recover). Indeed, there is a lot of commonality in the understanding of resilience in the three areas of psychology, ecology, and disaster relief.

In engineering the take on resilience is somewhat different. In fact, engineers more commonly use the term *robustness* with a connotation of "designed resilience." It differs from the other three uses in that it assumes bounded uncertainty—that is, the kinds and ranges of disturbances and shocks are known, and the system being built is designed to be robust in the face of these shocks. This view is now changing, and in chapter 5 we look at the emergence of what is being dubbed a "meta-robustness" approach. This sees a convergence of ideas about *resilience* as used in the other three domains.

In this book we present a definition and description of *resilience* that is being used commonly by scientists in many areas of inquiry. It is the capacity of a system to absorb disturbance and reorganize so as to retain essentially the same function, structure, and feedbacks—to have the same identity. Put more simply, resilience is the ability to cope with shocks and keep functioning in much the same kind of way.

A key word in this definition is *identity*. It emerged independently in ecological and psychosocial studies, and it is both important and useful because it imparts the idea that people, societies, ecosystems, and social-ecological systems can all exhibit quite a lot of variation, be subjected to disturbance and cope, without changing their "identity"—without becoming something else.

The following pages seek to present a simple overview of the essence of resilience thinking. If you can appreciate the following ten key points, you're in a good position to consider how you can move from thinking to practice.

1. The systems we are dealing with are self-organizing.
2. There are limits to a system's self-organizing capacity.
3. These systems have linked social, economic, and biophysical domains.
4. Self-organizing systems move through adaptive cycles.
5. Linked adaptive cycles function across multiple scales.
6. There are three related dimensions to resilience: specified resilience, general resilience, and transformability.
7. Working with resilience involves both adapting and transforming.
8. Maintaining or building resilience comes at a cost.
9. Resilience is not about knowing everything.
10. Resilience is not about not changing.

1. Self-Organizing Systems

First and foremost, resilience thinking requires that you recognize and appreciate that the systems we depend upon are complex adaptive systems. We use the more general term *self-organizing* systems because most people seem to grasp that more readily. Box 1 explains what the terms mean and the difference between being complex and being complicated.

All the things that most resource managers are interested in (e.g., farms, landscapes, and fishing grounds), but also things like your body, your family, and your business, are self-organizing systems. You can change bits of the system, but the system will then self-organize around this change. Other bits will change in response to your control. Sometimes you have a good idea about how the system will respond to your actions, sometimes it's difficult to predict, and sometimes the response comes as a complete surprise.

Most of the time the system can handle the changes it experiences, be they human management or some external disturbance such as a storm. By "handling it" we mean the system absorbs the disturbance, reorganizes, and keeps performing in the way it did—it retains its identity.

But sometimes the system can't cope with the change and begins behaving in some other (often undesirable) way. Sometimes a fishery crashes and doesn't come back when fishing pressure is removed. Sometimes an agricultural catchment becomes salinized as the water table rises and is no longer productive, even if the water table later drops. Even with the best intentions, our management sometimes turns our most precious ecosystems from valuable assets to expensive liabilities.

This often happens because our traditional approach to managing resources, which usually focuses on narrowly optimizing for some product (e.g., fish or timber or grain), fails to acknowledge the limits to predictability inherent in a self-organizing system. Don't worry if that sounds too technical; it makes sense when you work through a few of the concepts embedded in it.

2. Thresholds

There are limits to how much a self-organizing system can be changed and still recover. Beyond those limits it functions differently because some critical feedback process has changed. These limits are known as thresholds. When a self-organizing system crosses a threshold, it is

Box 1: Complex versus Complicated:
It's a Basic Difference

The word *complex* is used by all of us, usually when we are attempting to explain a difficult or tricky situation. For example, we might say, "This is a challenging and complex set of circumstances our nation is facing." In a resilience framework, the concepts of *complex* and *complex systems* carry particular meanings.

The three requirements for a complex adaptive system are

- It has components that are independent and interacting
- There is some selection process at work on those components and on the results of their interactions
- Variation and novelty are constantly being added to the system (through components changing over time or new ones coming in)

To understand *complexity* it's helpful to distinguish between *complex* and *complicated*.

The mechanism that drives an old-style clock is a set of tiny, intricate cogs and springs, often consisting of many pieces. This is a complicated machine and, to most people, a thing of wonder. However, the individual pieces are not independent of one another; rather, the movement of one depends on another in an unvarying way. Also, there is no selection process at work on the pieces, and these pieces don't change over time. It's a complicated machine but not a complex system.

Although a farm might produce just one item (e.g., wheat), the farm is far from simple. The farmer, the farming practices, the crop, the soil it grows on, and the market are all interacting and changing over time. This is a complex adaptive system.

Complex adaptive systems have emergent properties (i.e., their future states can't be predicted from the properties of their component parts). It's possible to control parts of the system for a time, but no one is in charge of the whole system. And because of all these features it's virtually impossible to keep it in some (optimal) state. Trying to do so initiates secondary feedback effects that can change and undermine the viability of that state.

The terms *self-organizing* and *self-regulating* are also used to describe systems with complex dynamics. Not all self-organizing systems have the emergent properties of complex adaptive systems, but self-organizing is an easier term to grasp, and so it is the one we tend to use.

said to have crossed into another "regime" of the system (also called a "stability domain" or "basin of attraction"). It now behaves in a different way—it has a different identity.

On coral reefs, for example, there is a threshold associated with nutrient levels. Plant nutrients find their way to coral reefs from fertilizers being used on the land. The nutrients wash off the land, eventually finding their way to waters around coral reefs. Nutrients stimulate the growth of algae. When the concentration of nutrients rises above a certain level, algae outcompete coral polyps for bare spaces on the reef. There is a critical level of nutrient concentration where this feedback effect on algae-coral competition takes place, and this is a threshold.

Below the nutrient-load threshold, corals predominate and coral polyps rapidly occupy any bare spaces created by disturbances. But if the reef crosses the nutrient threshold, algal growth overwhelms the young corals. It might be a storm that creates the bare space, but suddenly the system is behaving in a dramatically different way. It goes from a coral system to an algae system—it has a new identity; this change has major consequences for all the other organisms (including people) that depend on that reef.

In self-organizing systems you need to put the emphasis on thresholds because crossing them can come with huge consequences. Resilience practice is very much about thresholds—understanding them, determining where they might lie and what determines this, appreciating how you might deal with them, and very importantly, having the capacity to be able to deal with them.

Thresholds occur in ecosystems and in social systems. In social systems they are more often referred to as "tipping points." Tipping points might be changes in fashion, voting patterns, riot behavior, or markets.

Thresholds are often not easy to identify. Most variables in a system don't even have them; that is, considered on their own, the variables show a simple linear response to the change in underlying controlling variables and at no point exhibit a dramatic change in behavior (see figure 1a). For the variables that do have thresholds, it's important to know about them because they cause regime shifts. This means that once a threshold has been crossed, all the variables in the system are likely to undergo significant change. But, as we'll discuss, discovering where thresholds might lie is not easy.

And not all thresholds are the same. Sometimes you can cross a threshold but cross right back relatively easily. Water changes to ice when it crosses a temperature threshold of zero degrees Celsius, but it changes back to water when you raise the temperature above the threshold.

Sometimes there's a large step change when you cross a threshold, and then a similar large reverse change is experienced when you cross back, at the same point (see figure 1b). A common example of this is when some landscapes lose more than about 90 percent of their cover of native vegetation. Below this threshold there is a loss of a suite of native animal species from the landscape. However, provided they haven't been lost entirely from the whole region, restoring the landscape to more than 10 percent cover allows for their reestablishment (Radford et al. 2005).

Sometimes crossing a threshold involves a hysteretic effect. This is where the threshold you need to cross in order to return to the regime you've left is different from the threshold you crossed when you moved out of that regime in the first place. A couple of examples help to explain what we mean by this.

Many lake systems "flip" into a different regime when they get too much of the plant nutrient phosphorus (P). A small increase in P levels in the lake sediment pushes the system over a threshold, and it begins to behave very differently. Due to changes in P solubility under changing oxygen concentrations in the water, the amount of P in the water jumps much higher (it's very soluble under anaerobic conditions) and won't come down until P in the sediment is much lower. Algal growth is stimulated, and the lake goes from clear water to a regime of algal blooms and dead fish. This is shown in figure 1c.

Grassy rangelands that sometimes turn into shrub thickets offer another example. If grazing pressure reduces the amount of grass and causes shrub density to exceed some threshold amount, there then isn't enough grass to carry a fire. Fire kills many shrub species but not grass (which grows back from buds in its crowns below the soil surface). Without fire, the woody shrubs take over as the dominant vegetation. This further suppresses grass growth. The feedback from grass to shrubs via fire has changed, and even if grazing pressure is then reduced, the system stays in the woody shrub-dominated state for a very long time before shrubs die and the grass returns in sufficient amounts to allow fire to again play a role. And that delay might be enough to bankrupt the pastoralist. We look at rangelands in more detail in case study 1.

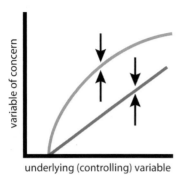

a. no threshold effect

variable of concern

underlying (controlling) variable

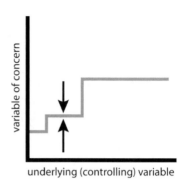

b. step change

variable of concern

underlying (controlling) variable

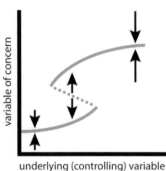

c. alternate stable state

variable of concern

underlying (controlling) variable

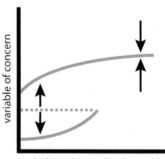

d. irreversible change

variable of concern

underlying (controlling) variable

Figure 1: Four Kinds of Threshold Effects

(a) No threshold effect: A dependent variable (e.g., crop production) changes steadily in response to a change in a controlling variable (e.g., rainfall). It can be a straight-line response but more commonly is an asymptotic curve. The arrows indicate that for any one level of the controlling variable there is only one stable amount of the variable of concern.

(b) Step change: Sometimes a small change in a controlling variable leads to a large step change in the variable of concern.

(c) Thresholds between alternate stable states: Sometimes a small increase in a controlling variable flips a system into a different stable state. Note: because the system has alternate states for some levels of the controlling variable, there are two values of the variable of concern.

(d) Irreversible threshold changes: Sometimes a small increase in a controlling variable flips a system into an alternate stable state from which there is no effective return. When saline groundwater rises to within around two meters of the surface (x-axis = rising level of groundwater), the salinity in the topsoil (y-axis) will suddenly jump as capillary action draws the groundwater to the surface. The groundwater might subsequently fall, but chemical and physical changes in the topsoil mean salinity levels in the soil will stay at high levels.

Sometimes this hysteretic effect is described as a "lag effect" because returning involves a delay. The word *hysteresis* comes from the Greek *husteros*, which means "late." However, it's more than just a delay. The pathway back is different from the pathway that took you over the threshold in the first place. Unless you dramatically reduce phosphorus levels in the lake or shrub levels on the rangelands, you don't return at all. In other words, it's not a matter of how much time passes (i.e., it's not a lag); it is a matter of the amount of the controlling variable. The hysteretic effect results in a system having two alternate stable states (or regimes) for the same amount of the controlling variable.

The lake and rangelands systems are able to return to their original states if the controlling variables (e.g., nutrient loads, shrub cover) are reduced to much lower levels than those that led to the change. For some systems, however, crossing thresholds represents a one-way trip. When saline groundwater reaches the surface of an agricultural landscape, it's effectively game over. The saline water will devastate crops and trees, but worse still, it will change the structure of the soil. Sodium disperses clay particles, making the soil "soapy" and sticky, greatly reducing water infiltration into the soil. This happens to such an extent that the salt will remain in the surface layers for a long time after the water table sinks (figure 1d). It will take large quantities of water to flush the salt out.

Not only are thresholds critical to understanding the behavior of self-organizing systems, they are the basic limits to your enterprise. To use the phrase in a recent analysis of global-scale thresholds (Rockström et al. 2009), they define the "safe operating space" of your system.

Thresholds Can Move

So, for a number of reasons, thresholds are difficult things to deal with: they come in different forms and they're often difficult to spot (that is, until you've crossed them, and then it can be too late).

As if that weren't enough, some thresholds can move because of other changes in the system. This means that resilience (the distance your system is away from a threshold) can increase or decrease. For all thresholds, including those that are fixed (the two-meter water table salinity threshold in figure 1d is an example), you need to know what determines their positions in order to manage resilience. The ones that can move are the hardest to analyze.

For example, consider the nutrient-load threshold in connection to coral reefs. The position of this threshold depends on how many herbivorous fish there are. Above a certain nutrient load, algal growth is favored over coral growth, so if any little shock opens up some space, algae occupy it, displacing coral. In places like the Caribbean, high levels of fishing pressure have removed fish groups that graze down algae, and in these situations the nutrient threshold that triggers a flip from a coral-dominated reef to an algae-dominated reef is lower than in places where lots of grazing fish are present (as on the Great Barrier Reef in Australia). As the fish that control algae disappear, the nutrient threshold allowing algae to take over gets lower and lower and is more likely to be crossed (resilience is diminished).

So, to recap, thresholds come in different forms, are often invisible, and can move. They can occur along biophysical variables like nutrients and plant cover, but they also exist in the social and economic domains of your system.

3. Domains Are Linked

Many of the problems associated with managing natural resources relate to the fact that our approaches don't acknowledge that we're dealing with systems that have linkages between the social, economic, and biophysical domains that make them up. Fisheries, for example, are often based on models of how many fish can be harvested over time, but the models focus only on our understanding of the biophysical domain—the dynamics of the fish population under various levels of harvesting—and quotas are set accordingly.

History has shown that these models based on expectations of optimum sustainable yield often lead to the collapse of a fishery, initially because they fail to acknowledge thresholds. But these failures are then exacerbated by the effects of linkages to the economic domain that were also not included in the model. Individual fishers carry large amounts of debt in the purchase of their boats and fishing equipment, and their need to service this debt leads to overharvesting of the fish resource. For some fishers it's a choice between losing their boat this season, because the bank forecloses on their loan, or overfishing the resource this season. With the latter choice, they can service the loan but they risk a fish-stock collapse in a later season.

Most people deal with the certain short-term threat and deal with the uncertain longer-term threat later.

Changes in one domain (e.g., debt levels in the economic domain) will often lead to changes in another (e.g., overharvesting in the biophysical, or stress in the social), and these then feed back to cause further changes in the first domain. This is one of the hallmarks of complexity; in self-organizing systems you can't understand one domain without understanding the connections with others and their feedback effects.

And those linkages are possibly most important when thresholds are crossed, because crossing one threshold can cause the crossing of other thresholds in other domains, forcing the system into a new (undesirable) regime. What's more, experience has shown that going over a series or cascade of thresholds (e.g., the crossing of a debt threshold in the economic domain that causes the crossing of a biophysical threshold that causes collapse of a fishery) leads your system into a highly resilient alternate regime. In other words, a cascading collapse is very hard to return from, and resilience isn't desirable when it means you can't escape a bad situation.

Understanding the interplay between thresholds and the linkages between domains is critical to understanding the behavior and resilience of self-organizing systems. Take a moment to consider case study 1 (after this chapter) on rangelands. It provides a convincing argument on why managers can't afford to ignore this interplay in their planning and management. It discusses many of the points made so far about self-organizing systems, linkages, and thresholds.

4. Adaptive Cycles

The next thing to appreciate is that the behavior of self-organizing systems changes over time, not because of external influences but through internal processes. The way that the components of the system interact causes the system to go through cycles in which the connections between its components tighten, loosen, and even break apart. As this happens, the capacity of the system to absorb disturbance (its resilience) also changes, as does the potential for people managing the system to make changes.

Think of a forest recovering after a fire or a new business set up to make and sell a new product. As they establish themselves, these systems undergo a period of rapid growth as they exploit new opportunities and available resources. In business, it's the entrepreneurs that can often do well and get ahead. In the forest, it's the fast-growing generalist species (the "weedy" species) that prosper because they can cope well with a bit of variability.

However, over time the capacity and potential for rapid growth diminish because the system is no longer operating in an "open field." Availability of resources is decreasing as the operating space gets filled; connections between players (or species) are increasing and becoming stronger. The fast growers in the forest are being displaced by the dominant trees (the stronger competitors but slower growers) that soak up all the available light and nutrients. The entrepreneurs in the business are being displaced by the accountants (bean counters) and the middle managers who have to improve productivity through increasing the scale of the operation and introducing ever-more efficiency, often by cutting out perceived redundancies. Risk taking is no longer encouraged, or even tolerated, and the system is much more conservative in its approach to business. The system enters a phase of "conservation" in which net production gets very small and size levels off; the forest or business is no longer getting bigger.

But just as rapid growth can't go on forever, so too this mature conservation phase inevitably comes to an end. The forest biomass builds to its maximum (climax) level and becomes ever-more inflexible, with nutrients locked up in heartwood. It is more and more prone to a disturbance such as fire or a pest outbreak. Or the business has grown so complex it can no longer steer its way through a changing economy or seize opportunities (like new technologies) as they arise. It takes a smaller and smaller disturbance to initiate a collapse—which inevitably happens.

Just as night follows day, forests and businesses (and political parties and civilizations) rise and fall. And when they fall, things become very uncertain. Connections between components that were once locked tight are torn apart. Economic, social, and biophysical capital (e.g., nutrients locked up in trees and dead organic matter in the forest, financial arrangements and operating units in the business) are released, and the equilibrium of the previous conservation phase disappears.

The release can be brutal for some, and it's always an uncertain time. Resources are lost (nutrients are leached out, money and people leave the enterprise), but it also opens the way for renewal in which a new order or new generation rises up. Often the new order is pretty much the same as the old order, but sometimes it is something dramatically different (e.g., Kodak went from film to cameras).

As renewal proceeds, a new order, a new "attractor" (potential equilibrium state), may emerge; connections begin to grow between the components of the system, and before you know it, you're back in a phase of rapid growth.

The cycle that has been discussed here for a forest or a business is described by ecologists as an adaptive cycle (Gunderson and Holling 2002) and is observed throughout a wide range of self-organizing systems. Its four phases are rapid growth, conservation, release, and reorganization (or renewal).

There are times in the cycle when there is greater leverage to change things, and there are times when effecting change is really difficult (like when things are in gridlock in the late conservation phase). And very importantly, the kinds of policy and management interventions appropriate in one phase don't work in others.

Taken as a whole, the adaptive cycle has two opposing modes: a development loop (the fore loop, or front loop) and a release and reorganization loop (or back loop) (see figure 2). The fore loop is characterized by stability, relative predictability, and conservation, and this enables the accumulation of capital (which is essential in human systems for well-being to increase).

The back loop, by contrast, is characterized by uncertainty, novelty, and experimentation. It's the time of greatest potential for the initiation of either destructive or creative change in the system. Most people prefer the certainty of the fore loop, but it's the back loop that revitalizes the system by releasing and recombining resources that were increasingly locked up in the conservation phase.

Given how unpleasant release and renewal can be, it's comforting to know that most systems spend most of their time in the fore loop, which is generally slow compared with the back loop. In fact, if you look around, you'll find that by far the majority of the systems you see will be in fore loop phases. The downside of this, in terms of our ability to deal with back loops, is that most of our research and nearly all of our management and policy development focus on fore loop behavior. Almost no research has been done on systems in their brief, chaotic, but critically important back loop periods.

LIBRARY OF ROWAN COLLEGE AT BURLINGTON COUNTY

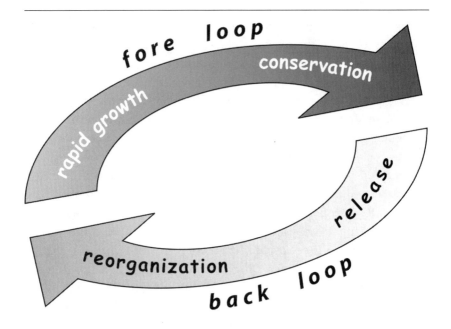

Figure 2: A Simple Representation of the Adaptive Cycle

In the fore loop the system is relatively predictable. The back loop is characterized by uncertainty, novelty, and experimentation. During the back loop there is a release and often a loss of all forms of capital. (From Walker and Salt 2006.)

The adaptive cycle is one valuable way to understand many self-organizing systems, but the cyclic pattern is not an absolute. There are many variations in human and natural systems. A rapid growth phase usually proceeds into a conservation phase, but it can also go directly into a release phase. A conservation phase usually moves at some point into a release phase, but it can (through small perturbations) move back toward a growth phase. Clever managers (of ecosystems or of organizations) often engineer this in order to prevent a large collapse in the late conservation phase. That is, they avoid a release phase at the scale of concern (the whole forest or the organization) by generating release and reorganization phases at lower scales and thereby prevent the development of a late conservation phase at the scale of concern.

Scale is very important to understanding resilience. So is the connection between scales, and that's what we discuss next.

5. Scales Are Linked

The adaptive cycle is a useful concept for understanding why a system is behaving in a certain way at a certain time, but it's only half the story. Self-organizing systems operate over a range of different scales of space and time, and each scale is going through its own adaptive cycle. What happens at one scale can have a profound influence on what's happening at scales above it and on the embedded scales below.

There's always one scale that is of particular interest to managers of a system. A farmer would be most concerned with what was happening at the scale of the farm. The prime minister in a government might be more focused on the scale of the nation.

In engaging with self-organizing systems, it's critical to acknowledge that you can't understand the focal scale (the thing that you're interested in) without appreciating the influence from the scales above and below—and often beyond that to larger and finer scales.

The notion that one scale of a system influences other scales is hardly a revelation. Anyone who has tried to change things (or taken an interest in history) will appreciate that change is never a simple or predictable process. The things that thwart our efforts to bring about change are, more often than not, happening at a higher scale and beyond our control.

For example, as we write this book, there is broad agreement that we have to do something about climate change and reduce carbon emissions, but it seems the best efforts of countless individuals, NGOs, and even some governments to bring about change aren't sufficient to shift the global status quo.

At finer scales, efforts by some farmers to conserve biodiversity or adopt new eco-friendly farming approaches receive little encouragement from government policy, which instead favors more traditional approaches to farming (focused on techno-efficiency). Change at the farm scale is prevented by the higher scale of government.

When you overlay the concept of linked adaptive cycles, the behavior of the whole system can sometimes make more sense. When the higher scale is in a conservation phase, change is difficult. Components are tightly connected, efficiency in doing the same things somewhat better is more important than experimentation, and the status quo rules. At other times, when the higher scale is in a growing, active phase, the efforts of farmers struggling to break a blockage are facilitated, not blocked, by the higher scales.

Finer scales will go through their own adaptive cycles, but as they emerge from a back loop, the higher scale may influence them to follow a path similar to the one they've just been on.

If a patch of forest goes through a release phase because it is damaged by fire, its restoration will be guided by the surrounding unburned forest, which will be providing the seeds and organisms that will return it to a forested state (and start it on a cycle similar to the previous cycle). Higher-scale dynamics constrain and can initiate and guide what's happening at the lower scale. The system as a whole has "memory."

A farm may go bankrupt, for example, because of inappropriate land policy, but when it rebuilds (or a new farmer steps in), it is still constrained by those same policies. It can be the other way around, too. Top-down influence can be positive as well as constraining and negative. Memory can be both good and bad.

But sometimes what's happening at a lower scale can tip what's happening at the higher scale. The term used for this is *revolt*, as reflected in the historical examples of social revolutions. History tells us that when the gap between rich and poor becomes too big for any length of time, the outcome can be a revolution (consider the French and Russian revolutions), beginning at small scales in local areas, in which the grass roots overthrow the ruling order. (Consider also what has happened in the Middle East.) It precipitates a back loop at a national scale. Social revolutions are usually horrible times of great unrest and insecurity, but they are also historical watersheds that reset and potentially revitalize a nation.

Revolt also happens in ecosystems. Once again, forests provide a good example. If only one or two patches of forest experience a pest outbreak, there is little change to the overall forest. As the affected patches regenerate, the memory of the higher scale imposes itself on the scale of the patch, and the patch regenerates back into forest. However, if the pest outbreak occurs in enough patches at the same time, it might cause a catastrophic collapse bringing down the whole forest. And if enough of the forest enters this back loop, it might signal a shift to a quite different cycle.

The process in which many cycles at the smaller scale become aligned is called synchronization, and it can lead to all sorts of problems, as it can massively destabilize higher scales and lead to revolt. A good example occurred in eastern North America in the years following the Second World War. The development of cheap pesticides gave forest managers the ability to suppress pest outbreaks at the patch level. They could suppress the pests but not eradicate them.

Then, instead of the forest consisting of a mosaic of patches in different stages of regeneration (some mature, some affected by a pest outbreak, and others regenerating), what they had was a forest all in the same state of development. The management had unwittingly synchronized all the patches, so they had made the whole forest (and the industry that depended on it) vulnerable to a massive pest outbreak, which would lead to a forest-scale back loop. The short-term solution was to apply ever greater quantities of pesticide. The longer-term solution was to explore ways of returning the forest to a patchwork mosaic. This example actually comes from the early work of C. S. (Buzz) Holling, who first formulated the idea of the adaptive cycle.

But what is the lesson for managers these days? If anything, it's this: It is all too easy and a bit of a trap to become focused on the scale in which you're interested. This scale is connected to and affected by what's happening at the scales above and the embedded components at the scales below, both in time and space. The linkages across scales play a major role in determining how the system at another (linked) scale is behaving. Sometimes the linkages and interactions seem obvious, but frequently they're only acknowledged in retrospect.

Ignoring cross-scale effects is one of the most common reasons for failures in natural resource management systems—particularly those aimed at optimizing production. The lesson is that you cannot understand or successfully manage a system—any system, but especially a social-ecological system—by focusing on only one scale.

Indeed, it's useful to imagine the system you're interested in as composed of a hierarchy of linked adaptive cycles operating at different scales (in both time and space). The structure and dynamics at each scale are driven by a small set of key processes and, in turn, it is this linked hierarchy of cycles—referred to as a panarchy—that governs the behavior of the whole system.

The term *panarchy* was originally coined by the ecologists Buzz Holling and Lance Gunderson. It describes the cross-scale and dynamic character of interactions between human and natural systems. It draws on the image of the Greek god Pan, a symbol of nature, and joins this with notions of hierarchies—cross-scale structures in natural and human systems.

6. Specified and General Resilience

Resilience thinking is the capacity to envisage your system as a self-organizing system with thresholds, linked domains, and cycles. Resilience practice is the capacity to work with the system in order to apply resilience thinking, to manage its resilience. Points 1–5 above were the basic building blocks of the resilience concept (self organization, thresholds, linked domains, and linked adaptive cycles across scales). Points 6 and 7 discuss aspects of resilience that inevitably arise when you begin to consider how you might apply the thinking.

To begin with, in terms of the current state of any system, there are two complementary aspects of resilience—specified and general.

Specified resilience, as its name suggests, is the resilience of some specified part of the system to a specified shock—a particular kind of disturbance. General resilience is the capacity of a system that allows it to absorb disturbances of all kinds, including novel, unforeseen ones, so that all parts of the system keep functioning as they have in the past.

One of the aims of resilience thinking is to identify possible thresholds beyond which the system will take on a new identity. In managing the system, you want to prevent it from crossing these thresholds, by controlling the state of the system or by influencing the position of the threshold (consider the example in case study 1). In doing so, you are working on the capacity of the system to deal with a specified threat. You are therefore attempting to increase its specified resilience.

But in a complex world there will always be disturbances and shocks that take you by surprise, that can't be envisaged or forecast. Air travel in Europe, for example, has become very elaborate and sophisticated in the options it offers travelers, whereas less focus (and investment) has been placed on sea and rail travel. It was a surprise when a volcanic eruption in 2010 shut down European airspace. It paralyzed international movements in the region (Folke et al. 2010) because sea and rail travel couldn't cope with the demand.

When you prepare your system for a specific disturbance, in a sense you're optimizing your capacity for a specific threat. In so doing, you may be eroding your system's general capacity to absorb other kinds of disturbances. This is reflected in the HOT (highly optimized tolerance) theory, which shows how systems that become very robust to frequent kinds of disturbances necessarily become fragile in relation to infrequent kinds (Carson and Doyle 2000).

In other words, there is a trade-off between specified and general resilience. Channeling all your efforts into one kind of resilience will reduce resilience in other ways. So it is necessary to consider both.

What are the things that enhance general resilience? Studies of a variety of social-ecological systems suggest diversity, openness, reserves, tightness of feedbacks, modularity, and redundancy are all important characteristics of systems with high levels of general resilience. Systems with rigid, efficiency-driven, top-down control and management, on the other hand, have little adaptive capacity and often have low levels of general resilience. Examples of this in business are just-in-time marketing and production systems in which all perceived redundancy is eliminated from the production and marketing processes. In these systems everything is locked into being highly efficient. Stockpiling and storage (and all the transaction costs that this entails) are eliminated. These are efficient systems that work wonderfully until some disturbance hits any part of the supply chain, and then they simply stop working, sometimes with dire consequences for the business.

The way a system is being governed also affects general resilience, as do aspects such as leadership, trust, and social networks (collectively sometimes referred to as social capital).

It's important to realize that both kinds of resilience come into play in determining the ability of a system to absorb a shock without crossing a threshold. The likelihood of that happening depends on two things: (1) how far the current state of the system is from a threshold and (2) how much the system is changed (disturbed) by the shock.

The first is basically about knowing where thresholds might be and where the system is in relation to them (specified resilience).

The second is largely about the system's capacity to manage the disturbance—to prevent the state of the system from reaching the threshold. It is about the resources that can be brought into play and the other attributes described above that provide general resilience.

How you might approach assessing a system's specified and general resilience is discussed in chapter 3.

7. Adapting and Transforming

Resilience is often portrayed as a universally good thing, and government rhetoric often tells us that resilient economies, communities, and landscapes are something to strive for. But this assumes that the economy, community, and landscape being discussed are in a desirable state that you want to maintain. What if the economy is in recession, well-being in the community is low, or the landscape is degraded?

Resilience per se is not good or bad. Undesirable states of systems can be very resilient. Rangelands choked by woody weeds, salinized catchments, and military dictatorships can all be highly undesirable system states that are also highly resilient. Sometimes it isn't worth the effort to keep doing what you have been doing. Sometimes it's not even possible, and yet we often persist anyway. Under these conditions, however, persisting just makes things worse.

When it comes to managing resilience, you can aim to maintain the identity of your system; in other words, you can adapt and build up the resilience of the current state of your system. Or if the system is in an undesirable state, you can try to get back into the desirable state by reducing the resilience of the undesirable state. But sometimes that is impossible. When that is the case, you can aim not to adapt but to reimagine your system as something else, to transform—in other words, become a different system.

Adaptability is the capacity of a social-ecological system to manage resilience—to avoid crossing thresholds, or to engineer a crossing to get back into a desired regime, or to move thresholds to create a larger safe operating space.

Transformability is the capacity of a system to become a different system, to create a new way of making a living. An example comes from southeastern Zimbabwe where, in the 1980s, ranchers transformed their cattle ranches to game-hunting and safari parks when the livestock industry proved unviable. Transformability is discussed in more detail in chapter 4.

On the surface, it may appear there's a tension between adapting and transforming. Should you adapt or transform? But the tension is resolved when you consider the system at multiple scales, because making the system resilient at a regional scale, for example, may require transformational changes at lower scales (Folke et al. 2010). Adapting and transforming are actually complementary processes, and adaptability and transformability are complementary attributes of a resilient system.

A good example of this in practice is the current proposed change to water allocation in Australia's Murray Darling Basin. Huge cuts in many subcatchments have been identified as necessary for the basin as a whole to continue functioning. To retain its identity as an agricultural region, it will require transformational changes in a number of its irrigation areas, from irrigation farming to some other kind of agriculture.

Sometimes the system won't change easily. The process of transforming is never without pain. However, if transformation needs to take place, it's better to do it sooner than later. The costs of delay can be extremely high. For example, faced with the growing specter of climate change, many call for a move to a low-carbon economy. Most economists and climate scientists believe the enormous costs associated with delaying this transition make the case for early transformation compelling.

Furthermore, as discussed later, the pace of both environmental and social changes in the world suggest that transformational change may need to be a continuing process, rather than a one-off or periodic major change.

The three ingredients necessary for transformation are

- The preparedness for change (as opposed to a state of denial)
- Having the options for change (possible new "trajectories")
- The capacity to change

These are all discussed in chapters 3 and 4.

8. Resilience Comes at a Cost

Building resilience isn't free; it comes with both the direct costs of the actions you take and the indirect costs of opportunities lost by not using your resources in some other way.

Enhancing the resilience of a system usually involves reducing efficiency, staying away from maximum yield states, maintaining reserves, and so forth. When this happens in response to a specified threat, it's theoretically possible to measure the cost (and the benefit) of what you do. For example, to avoid a buildup of nutrient around coral reefs, managers may use a range of strategies, and each of these has its own cost. The total cost can then be compared with the risk of the system crossing a threshold into a new regime (e.g., coral reefs becoming algal reefs), and decision makers can then weigh the cost of stopping this from happening against the cost of not stopping it.

In this manner it's possible to estimate and assess specified resilience. But the same does not apply to general resilience, because no specific alternate states of the system, associated with specific costs, are being compared. So how do you assess the trade-offs relating to general resilience? The basic approach involves assessing the sources of general resilience in your system (diversity, reserves, modularity, and so forth), monitoring them to see if they are in decline, and then determining whether this decline is a problem. This is discussed in chapter 3. The inability to assess the costs is a reason why general resilience is often allowed to decline.

9. Not Everything Is Important

In our discussions so far on the essence of resilience (points 1–8), we have tried to set out what resilience is. It's an emergent property of a self-organizing system. It's about adaptive cycles, linked scales and domains, specified and general resilience, and two complementary strategies of intervention (adaptation and transformation). It's all the stuff that is supposed to give you a handle on complex adaptive systems.

Got it? Maybe you have. However, for most people, on first exposure this is all quite a mouthful, possibly too big a horse pill of information to digest in one sitting. If you are struggling, don't get frustrated, because it does become clearer as you consider this framework with case studies and in interpreting the systems that you're interested in.

Resilience thinking isn't simple, but it is simpler than it first appears. First impressions are that to understand your system—your complex adaptive system—you have to know everything about everything. Everything seems connected to everything else, and until you have a comprehensive knowledge of everything (i.e., thresholds, scales, cycles, feedbacks, and domains), there's nothing you can do.

You do have to know quite a bit about your system. You do have to develop an idea about its thresholds, scales, cycles, feedbacks, and domains. But everything is not connected to everything else, and you don't have to know everything about everything.

A key phrase in resilience thinking is *requisite simplicity*—as simple as possible, but not too simple. Resilience thinking actually aims to help you identify the minimum but sufficient information you need to effectively manage your system for the values that you hold to be important.

Indeed, one important insight that has arisen from decades of research on the resilience of social-ecological systems is that the important changes in the system, the ones that can constrain and redefine the futures of regions and whole communities, are determined by a small set of three to five key variables at any one scale. This is known as the "rule of hand," and it says that to understand significant change in systems, it is important to identify this small set of variables.

So it's *not* about making things more complicated. It *is* about enabling you to engage with complexity and focus on what's important. Resilience thinking is a problem-framing approach to your system that seeks to help you decide what's important for the sustainability of the things you value, that you should be focusing on.

10. It's *Not* about Not Changing

Our final point is about something that resilience isn't. It relates to a misperception that sometimes arises, that being resilient is about keeping things the same or bouncing back to exactly the same condition.

Resilience is *not* about not changing. This sounds a little paradoxical, but it's an important point that is worth taking a moment to appreciate.

Consider again the basic definition of *resilience*. It's the capacity of a system to absorb disturbance and reorganize so as to retain essentially the same function, structure, and feedbacks—to have the same identity. Sometimes people read this as "staying the same" and think that resilience is all about keeping things exactly as they are. However, being resilient *requires* changing within limits—in fact, probing those limits.

A resilient coral reef is one that can reassemble after the battering of a hurricane. A resilient rangeland is one that recovers its productivity after a fire or a drought. A resilient forest is one that can grow back to the same kind of forest after a pest outbreak. A resilient business is one that can absorb a market shock and return to profitability. In each of these situations, the basic identity of the system stays the same though each system is changing all the time—by changing, it enhances its resilience.

Holding a system in exactly the same condition erodes resilience because the capacity to absorb disturbance is based on the system's history of dealing with disturbances. So, for example, if you prevent savannas from occasional burning (through grazing management or fire control), they lose their resilience to fire—they become vulnerable to it. A savanna that is never burned eventually loses its species that are adapted to fire, and when a fire eventually (inevitably) occurs, it has devastating results. The only way to maintain the resilience of a savanna to fire is to allow or cause that savanna to periodically burn.

Protect a company or an industry from overseas competition, and it loses the capacity to compete on overseas markets and becomes more vulnerable to changes in markets.

Staying exactly the same is actually a prescription for the loss of resilience because the system loses its capacity to deal with change and disturbance. Traditional approaches to resource management, based on ideas of optimal sustainable yield, often fall into this trap. They attempt to hold a system in a configuration that achieves the greatest productivity, without acknowledging the dynamic nature of the system they are attempting to control.

If you consider the range of different states that your system can exist in as a basin of attraction, the act of keeping the system the same, in one particular state, causes that basin of attraction to shrink. The distance to thresholds beyond which lie other basins of attraction (other regimes) is reduced, and it takes a smaller and smaller disturbance to shift your system across those thresholds.

A resilience approach is about acknowledging change, embracing and working with it. Resilience thinking is structured around the acceptance of disturbance, even the generation of disturbance, to give a system a wide operating space.

Indeed, in the words of our colleague Steve Carpenter, resilience thinking is really about changing in order not to change.

From Thinking to Practice

So there it is, the essence of resilience thinking in ten points. Our "essential guide" is simply our interpretation of what you need to have some idea of before attempting to move into practice. You'll find similar catalogs of resilience concepts in different forms in different places. Resilience thinking is multifaceted and you should engage with it in different ways to get the most out of it.

However, with our ten points laid down, let's now see what we need to do to put the concept into practice. We're proposing three stages. They loosely follow the approach set out in the workbook of the Resilience Alliance (2010). However, once again, to make it work, you have to make it your own.

The three stages involve *describing* the system (chapter 2), *analyzing* its dynamics and what this means (chapter 3), and then *deciding* what you can do about it (chapter 4).

But our discussion on what resilience practice is goes beyond this. In chapter 5 we discuss what resilience means and how it is used in other specialist areas, such as disaster relief and engineering. In chapter 6 we consider resilience at a global scale and what a resilient world would look like.

Key Points for Resilience Practice

- Resilience is a dynamic property of a system, and managing it requires a dynamic and adaptive approach. There are many ways of putting resilience science into practice.
- The basics of resilience thinking involve appreciating a system's thresholds, domains, and linked adaptive cycles.
- Resilience thinking is about understanding requisite simplicity. What, essentially, do you need to know about your system to keep it sustainable?
- Specified resilience is how far the current state of a system is from a threshold. General resilience is the system's capacity to manage a disturbance and prevent the state of the system from reaching a threshold.

CASE STUDY 1

Thresholds on the Range:

A Safe Operating Space for Grazing Enterprises

R angelands are places where humans graze animals for meat and fiber. At their simplest, they can be pictured as expanses of grassy woodlands or grasslands with shrubs, managed by pastoralists who graze animals on them. They are a foreign world to your average city slicker, but rangelands supply an important proportion of the world's protein (especially to the developing world). They have also experienced significant degradation over many decades.

There is a range of rangelands. They are all semiarid but some get more rain than others. At the wetter end of the spectrum the problem of bush encroachment, or "woody weeds," has occurred in rangelands all around the world. The nub of the problem is competition between shrubs and grass for water, an interaction that is mediated by grazing and fire.

Shrubs are slower growing than grasses, but their deeper roots make them stronger competitors. However, when there is a lot of grass around, savannas burn, and they burn quite often. Indeed, rangelands evolved with fire and it's a necessary part of their environment. Under (human-controlled) grazing by livestock, fires are less frequent than in equivalent natural ecosystems, but fire is still used by managers because it kills the shrubs but not the grass.

As shrub density increases, the amount of grass that can grow in a season decreases. Eventually, above some critical amount of shrubs, there is too little grass to carry a fire in the dry season, even if there has been no grazing. Over this threshold the rangeland "flips" from a

Images 1 and 2

Two alternate states of a rangeland system in Australia: an
open, grassy state (top) and a
dense, shrubby state (left).
(Photos: D. Tongway.)

grassy to a shrubby state, and it can take decades before shrubs die and grass can reestablish (Anderies et al. 2002). Consider images 1 and 2.

At the drier end of the spectrum of rangelands, shrubbiness is less of a problem. What is important in these drier rangelands is the loss of grass cover and desertification. This involves a biophysical threshold involving the amount of grass cover on the soil. Infiltration of water into soil is around ten times higher under grass and litter than in bare soil (on all but very sandy soils). There is a critical amount of grass cover below which water runs off the surface of the soil rather than penetrating the soil, and thus there is less water in the soil available for grass growth. Below some critical amount of grass cover, the grassland cannot restore itself, even if the livestock are removed. It takes an exceptional series of events, or some disturbance of the soil surface, to get sufficient water into the soil to get the grass cover back above the threshold.

The savannas of northern Australia are rangelands where this loss of infiltration is important (even though they lie at the moister end of the rangelands). These savannas are used to graze beef cattle. It is open eucalyptus woodland with an understory dominated by native perennial grasses. Pastoral properties are large, ranging in size from a few thousand to over a million hectares. Managers rely on few inputs and have only modest outputs per unit of land. A highly variable climate means that grass supply varies greatly from year to year, and this, together with relatively infertile soils, creates an environment that is susceptible to overgrazing (Ash et al. 1997).

A similar story occurs in the much drier Sahelian rangelands in Niger, a landlocked country in western Africa, and in Kenya's Masai land. In these regions household operations are much smaller than in the Australian station properties. When grazing pressure results in declining grass cover, a point is reached where desertification occurs.

Ecologists familiar with the dynamics of these rangelands will know that the above description is an oversimplification, so we need to elaborate a little. The amount of grass cover, and the critical threshold level, is strongly influenced by the spatial patterning of the vegetation and the consequent spatial dynamics of the landscape hydrology. The vegetation self-organizes into a two-phase pattern of (larger) runoff zones and (smaller) "run-on" zones. It was first described in the 1930s by French ecologists in west Africa who described the patterns they saw on aerial photographs as *brousse tigre* (tiger bush)—the strips of vegetation separating the relatively wider, bare runoff zones resembling tiger stripes. It has since been observed also in America and Australia, and its dynamics are well described by Ludwig and Tongway (1995) in Australia and by Rietkerk and colleagues (2004) in west Africa. In the African studies it has been shown how the changes in the pattern can identify the critical point at which the whole rangeland shifts into a degraded state.

This is far from just an academic observation. The dynamic spatial patterning makes for a resilient rangeland across a range of grazing intensities. However, below some critical level of cover, the surface flow of water changes from a dispersed flow that is intercepted by the vegetated patches to a channelized flow that forms erosion gulleys. Once this happens there is a net loss of water from the landscape, as a whole, leading to desertification. There is a threshold level of grazing intensity above which a rangeland suffers progressive net loss of the rainfall it receives.

It's relatively easy to appreciate a threshold of grass cover operating in a dry rangeland. A pastoralist puts too many cows onto the land, they overgraze the grass until the landscape hydrology changes, the rangeland becomes degraded, and the enterprise becomes unproductive. It's not an uncommon occurrence.

It would appear that this has everything to do with a biophysical variable (grass cover) and the manager simply needs to know how many head of cattle to run at different times to stay above that threshold level of grass. Of course, there's nothing simple about that calculation, and good pastoral managers need many years of experience to know what a "sustainable" stocking rate is.

However, you don't have to be an experienced farmer to know that this isn't just about percentage cover of grass. It's also just as much about income, profit, and debt: that's why the cattle are out there in the first place—to generate a livelihood. Just as there are thresholds in the biophysical domain of the system, so too are there thresholds in the socioeconomic domain, and they are linked.

In one analysis of grazing enterprises on savannas in northern Australia (quoted in Fernandez et al. 2002), it was reported that a threshold exists around 60 percent perennial grass cover. If the cover drops below this, the system moves into a degraded condition, and it takes a considerable intervention or an unusual sequence of favorable seasons to return it to a productive state.

But it's not just this biophysical threshold that grazing managers need to be wary of. Economic analyses of this same system suggest that when equity ratios (debt:capital value of farm, expressed as a percentage) drop below about 80 percent, debts become extremely difficult to service. In other words, the grazing manager has no choice but to run as many cattle as possible to maximize short-term income (to service the loan) even if that runs down the long-term capital of the farm.

This same situation applies to many rangelands, and the Australian example is one of several in an analysis of how desertification can occur in semiarid regions. The all-too-common occurrence of poverty traps in Sahelian rangelands is part of this same syndrome. The pattern of change is illustrated in figure 3.

The state of the rangeland depicted in figure 3 is defined by two variables: grass cover (acknowledging the complexity in earlier discussion on spatial dynamics) and income:debt (or capital:debt) ratio. As long as the enterprise doesn't cross the biophysical threshold (BT) or the economic threshold (ET), it can continue to operate safely. Cross one of those thresholds, however, and the system begins to move toward the other one, and therefore a degraded condition. The further it moves beyond the threshold, the harder it will be to return.

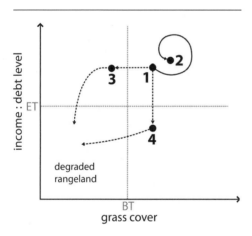

Figure 3: Two Thresholds in a Rangelands System

If the system (at position 1) stays above the biophysical and economic thresholds (e.g., moving from 1 to 2, or anywhere in this quadrant), the enterprise can continue to operate safely. If, however, grass cover drops below the biophysical threshold (from 1 to 3) or the debt levels push it under the economic threshold (from 1 to 4), then changed feedbacks in the system drive it into a degraded condition.

The important point to note is that crossing one threshold strongly increases the chance that the other will be crossed. The domains and the thresholds are linked. Run too many cattle and you might cross the biophysical threshold because the enterprise loses productivity, and debt levels are then likely to rise. On the other hand, cross the economic threshold and you'll be forced to run too many cattle to try to service the debt, and then you'll be much more likely to cross the grass threshold. Same endgame.

Though this is a conceptual representation of the rangeland system, it makes it easy to envisage what a "safe operating space" consists of. It doesn't matter if your system moves around within this space (as indicated by movement around state 1 in figure 3), as is likely from year to year. As long as you don't cross either the biophysical or economic threshold, the system will continue to work for you rather than against you.

And just as the top right quadrant of this space might be considered a safe operating space, the bottom left-hand space might be thought of as a trap—a doubly defined desertification trap. And it's a trap that is usually difficult or impossible to get out of. Very often the only option in this situation is transformation.

Of course, identifying exactly where these thresholds lie isn't easy (and how you approach this is discussed in chapter 3), but you also need to keep in mind that these thresholds aren't fixed; they can move (see figure 4).

In the Australian example, one way you might move the economic threshold is by improving the knowledge behind your management decisions. A good example of this is more effective medium-term weather forecasting (something that is becoming available in many developed countries). If there is greater certainty that the following season will be favorable, a farmer can increase stock numbers well in advance and take advantage of the good times and still stay away from the ecological threshold. Likewise, advanced notice of a poor season will enable reducing stock in time. In the longer term, better forecasting increases average incomes, meaning the farmer can service a higher debt. In other words, the economic threshold is raised, thereby increasing the safe operating space of the enterprise.

In the Sahelian case, from discussion among those familiar with the region, it seems that one way of moving the threshold is by having more off-farm income. The less a household's welfare depends on the income from rangeland, the higher the debt level it can manage and still be able to recover.

In regard to the ecological threshold, one way of increasing resilience of the grass cover is to keep as much of it in the form of

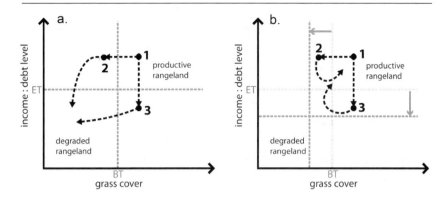

Figure 4: Moving Thresholds Can Change the Size
of the Safe Operating Space

Thresholds are not fixed. If thresholds can be lowered, it's possible the trajectory of the system can be shifted. In both (a) and (b) the system has moved the same distance: from 1 to 2 in terms of loss of grass cover and from 1 to 3 in terms of increasing levels of debt. In (a) this has resulted in the crossing of biophysical and economic thresholds, and the system tracks into a degraded condition. In situation (b) the thresholds have moved, the system no longer crosses them, the feedbacks don't change, and the system remains in a safe operating space.

perennial grasses (as opposed to annual species) as possible. Perennial grasses have much stronger root systems and promote higher infiltration than annual grasses for the same amount of cover. And the production of perennial grass varies much less from year to year in response to rainfall variation. So with perennial grasses the threshold is at a lower level of grass cover, and since the state of the system varies much less, there is less chance that it will be pushed across the threshold in a bad year.

However, just as it is possible to increase your safe operating space, it's equally possible for it to shrink. For example, if interest rates are raised, suddenly the debt threshold might shift up (i.e., less debt can be tolerated). Or if there's an invasive plant species, it's possible the amount of grass cover you need might go up.

Either way, if the farm is operating close to a threshold, it might, without warning, find itself on the wrong side of the threshold because of factors beyond the farmer's control—factors, for example, arising at a higher scale.

Lessons for Resilience Practice

Managing a resilient grazing enterprise on the rangelands is about keeping the enterprise in a safe operating space bounded by thresholds beyond which the enterprise will become unsustainable. It is not about controlling the enterprise to keep it at some single point of optimum economic return or level of livelihood production. Resilience practice is partly about attempting to envisage where thresholds might lie and treating them in an adaptive, learning manner. It's important to keep in mind that these thresholds occur in different domains (biophysical, economic, and social) and that they interact. Crossing one influences the likelihood of crossing another. It's also important to appreciate that thresholds can shift. The size of a system's safe operating space reflects its resilience.

2

Describing the System

A resilience assessment begins by bringing together the stakeholders—the people with an interest or a stake in the system. Many people find the term *stakeholders* a bit repellent because it arose out of management speak, but we can't think of a better one. The first stage is to work with these stakeholders to determine what the components are that make up their system and how they are connected.

Experience from a succession of workshops on assessing resilience across a range of agroecological regions suggests that an appropriate process for describing the system is to work through five steps. Although we present them in a sequence, they all overlap and can be dealt with in any order. The five steps are

- Scales (bounding the system)
- People and governance (the players, power, and rules)
- The resilience *of what* (values and issues)
- The resilience *to what* (disturbances)
- Drivers and trends (history and futures)

It's important not to get bogged down in any topic but to keep moving between the categories, because they inform each other. It's likely that while discussing one category you'll generate insights into others, and you're encouraged to jump from topic to topic adding to your lists, notes, and maps as ideas come up.

What we say now seems so obvious that it doesn't need saying—but it needs to be stressed. You could do this exercise by yourself, but by making it a group exercise you're engaging the perspectives and experiences of different people. We all have different insights into our systems, and system descriptions are stronger and more complete for including multiple perspectives. But not only that, if you don't engage the critical set of groups who have a legitimate stake in what happens, it puts into question the legitimacy of the assessment, lowering the likelihood of its acceptance and implementation. Identifying and engaging the critical set of stakeholders is one of the hardest things to get in place for a resilience assessment.

This question of legitimacy is a difficult one. The development of a social-ecological inventory (in the sense intended by Schultz et al. 2007) has proved a useful way to identify and engage local stakeholder groups, the "stewardship" groups who have a strong interest in what happens. However, as described by Schultz and colleagues, it is quite a lengthy process. It can be done iteratively and therefore not be too cumbersome, but experience suggests that time for busy people is a strong limiting factor.

While the major government agencies and industry groups are unlikely to be overlooked, local stewardship (and other interest) groups sometimes are. The social-ecological inventory described by Schultz and colleagues is designed to engage these groups, whose omission will threaten the legitimacy of the assessment. They used interviews with key people to identify such groups and individuals, plus an "extensive review of other information sources, including project proposals, progress reports, notes, maps, correspondence, internet sites and newspaper clippings."

But it's not only the stakeholder groups that need to be considered. You also need to consider who has the knowledge that needs to be included. Brown (2010) emphasizes the importance of including multiple knowledge groups in collective learning programs and identifies five of them—individual (own experience), community, specialist, organizational, and holistic. Having all of them involved in a resilience assessment increases the likelihood of a successful outcome.

The more preparation put in at the beginning, the higher the likelihood of acceptance. Shortcutting this engagement process is not a good idea.

Another important point to be made at the start is that this is not a test, and there are no exact answers. Indeed, it's unlikely your first efforts to describe the system will survive without significant modification. It's an iterative process and each time you ask the questions in each section, you'll be generating new information and insights that will feed back and alter what you've already recorded.

How Significant? How Long? How Much?

Our experience is that early efforts at describing the system yield a wealth of detail that subsequently is found not to be necessary; but that's not a problem. When you start out, no one is sure what's significant and what's not. That becomes clear over time and after multiple iterations. As the practice continues, the aim is to reduce the details to the necessary minimum (and what that might be is changing all the time too, so you have to constantly revisit it).

What are the significant components and interactions in your system that need to be taken into account? Some people find this part frustrating because on first inspection it seems that everything must be included, assuming everything is connected to everything else. In fact, the aim of resilience practice is actually the opposite: what's the minimum, but sufficient, information we need to incorporate in our understanding—our models—to make robust decisions about planning and management? Remember the phrase *requisite simplicity*; it needs to be kept constantly in mind.

How long should it take to describe your system? Aim for somewhere between a day and a year. In case study 3, we outline a process undertaken in Australian agricultural catchments that has so far involved several sessions, with more planned. The difficulty of bringing together a bunch of managers and decision makers is that you'll realistically only be given one or even only part of a day to undertake the description in a formal sense. However, the process of describing the system is really ongoing because it takes time to collect, assimilate, and reflect on the collected knowledge of the many people interested. The more information you can have assembled before the group process begins (maps, statistics), the better.

It is really important for all involved to have—to collectively develop—a common understanding of what the system is, and what the big issues are. It may take several passes, but make sure the description of the system is done well.

It's also an ongoing process because you are dealing with a self-organizing system, and self-organizing systems are complex, dynamic, full of surprises, and uncontrollable. In attempting to describe it, you're building and refining the stakeholders' own mental model of what it is, but this model will only ever be a rough approximation of what it actually is, and it'll never be complete. It becomes, rather, a way of thinking about the system.

Describing the system (stage 1) sets the scene for assessing its resilience (stage 2; see chapter 3) and then managing that resilience (stage 3; see chapter 4). However, assessing and managing inevitably generate a whole new set of insights that better inform your original description. Once again, it's an iterative process. And, once again, don't worry too much about getting it "right" first time around (and that applies to the second and third time, and to iterations after that).

1. Scales

Bring a group of people together and ask them to describe "their system," and you're bound to start a lively debate as they emphasize what's important to them. This can be resolved to a considerable extent by getting the group to agree on the critical scales at which the system functions (some will have different scales in mind) and appreciating that you cannot understand and define a system at only one scale.

An important first step, therefore, involves getting the people involved to set out an understanding of what their system is in spatial terms—the different scales at which it operates. We need to bound the system (and the issues we are concerned about), and the first scale we attempt to define is the one we are most interested in—the focal scale.

The way people view and govern natural resources (crops, forests, fisheries) varies around the world. How would you frame the area that encompasses what's important to you? In agricultural catchments it seems logical to describe the focal scale as the catchment itself, but at resilience workshops with catchment managers it often turns out that they see the areas of most interest (the scales they are more interested in and can better focus on) as components of the catchment where similar activities are being carried out. For example, one component might be dryland grain production in one part of the catchment, as opposed to irrigated horticulture in another part. Both boundings, the catchment as a whole and parts of the catchment, are equally valid. It all depends on the group of stakeholders and the issues that have initiated the assessment. Either can be appropriate.

Coastal managers grapple with how to define the "coast"—how far out to sea and how far inland? One option might be the seaward influence of the land and the landward influence of the sea. The Seri fishers of Mexico (as described in case study 4) might logically focus on the Infiernillo Channel, since it bounds their entire fishery, but they might also like to include in their focal scale the land area where interactions between it and the sea are central to their concerns.

River managers might identify a section of river within which similar activities are taking place, or an area of river and land that directly influences the issues they are most interested in.

Self-organizing systems operate over multiple scales. What happens at your focal scale will be affected by what happens at smaller (embedded) scales and larger scales. If the focal scale is a wheat-cropping area in part of a catchment, your enterprise will be affected by values and practices at the farm scale, and also by the rules and regulations, and by activities like markets at the catchment, region, state, and country scales.

To understand resilience you need to know what is happening at these other scales and understand how connections between scales might be influencing what's happening at your focal scale. Natural resource management failures often come about because decisions are being made at one scale without consideration of the connections with, and feedbacks from, other scales. At this point, don't worry too much about the interactions between scales. They become clear with subsequent work. To start with, you just need to determine what the significant scales are.

Considerations of scale need to include both the biophysical spatial dimensions (e.g., farm, subcatchment, catchment, region) and the scales of the social domain. A common problem in natural resource management is that the *biophysical* boundaries, often used as the basis for land-use planning, are often quite different from the *social* "catchment." Where this mismatch in the boundaries is significant, it causes inappropriate resource use and social discontent, so try and identify whether any spatial mismatches exist between how your social systems function and how the biophysical systems function.

The scales that different layers of governance operate at also need to be defined. What resource sectors are present in your focal scale?

For example, are they dryland farming, irrigated farming, biodiversity conservation, water flow regulation (weirs and dams), forestry, mining? From where do the people in these sectors get the inputs they need, and where are their markets? Consideration of sector activities helps define important scales and interactions across them.

Outputs from this discussion on scales might include a series of maps and lists that pool the ideas of the participants. The maps can be as rough or as detailed as you like. The lists are for reference as the description moves on to other aspects of your system.

2. People and Governance

Don't get bogged down on scales. You need to move on to the session on the big issues, because getting those clear not only helps clarify scales, but it starts to identify the "resilience of what." However, before you can do that, you need to know who the players are and how the governance system works. Therefore, before getting into listing and dissecting the big issues, it's important to consider the people and the rules, the governance in and of the system. Then, with this information, consider the big issues.

As we have already emphasized, getting the right people involved at the start, and being clear about who controls what and who has legitimate interests, is important. You may want to toggle between the two questions, Who? and What?

Governance includes all aspects of rules and regulations that determine what and how people operate in the system, as well as the different kinds of institutions that influence or determine how people behave. There are many rules at play, and it is all too easy to get bogged down in their details. The intention is not to list everything but rather to consider governance under a number of headings and to see if there are any issues that are causing problems. So, don't feel you have to answer all the following questions. Use them simply as a guide to help you identify issues worth noting.

The Users of the System

What are the "user groups" (sectors) and what are their "rights" or entitlements—especially their property and access rights? (This will have been dealt with to at least some extent in the initial step of identifying and engaging the relevant stakeholder groups.)

Property rights are often an issue. What rights do people have to access or control resources? Are property rights and access rights clear and agreed to by all, or are rights a contentious issue? How do the different kinds of tenure conflict with or complement each other, and is their juxtaposition a factor in this?

Who are the "secondary" users—suppliers, repair shops, and so forth? How significant are they?

Governance

Who controls resource use and regulations at each relevant scale? Are there problems in the relationships between the control agencies? Do the problems hinder or otherwise influence appropriate resource use? Are the objectives of the agencies compatible, or do they give rise to conflicts? How much overlap is there? You need to consider

- Government departments and relevant legislation, at different scales and levels of government, down to local government
- Industry organizations (co-ops, irrigation companies)
- Formal and informal institutions, such as Landcare groups in Australia and fishing co-ops in Sweden, local conservation societies, lobby groups, recreational associations, and so forth

The social-ecological inventory exercise mentioned above is one way to identify these groups. How effective are social networks and what role are they playing (or could they play) in learning and management?

These questions about users and governance will play a role when you come to assessing general resilience.

3. The Resilience of What? (What Do People Value and What Are the Big Issues?)

What *is* the system? What does it consist of, in terms of the biophysical (vegetation types, water bodies, agroecosystems) and social (culture, values) components? What are its connections to and inputs from surrounding areas and higher scales (rivers, infrastructure, markets)? You have already discussed this in part in steps 1 and 2. In this step you're attempting to be more explicit about what is important in the system.

The key questions are, What is it about the system that you want to be resilient? What do people value in, and want out of, the system? And what are the big issues that concern them?

A useful way to approach these questions is to use an ecosystem goods-and-services framework, as developed in the international program known as the Millennium Ecosystem Assessment (MEA 2005; see box 2). It's useful because it makes people think about not only the things they directly benefit from, like crops and water, but also the indirect things such as flood control and pollination services for their crops.

In systems where primary production or harvesting dominates, a common concern is the economic health of the enterprise. Wheat farmers want good harvests of wheat attracting high prices at markets. Fishing industries want good catches and strong prices; and the same applies to most primary industries. But some stakeholders place a high value on, and are most concerned about, the biodiversity in the system. Biodiversity is not a product from the system; rather, it's a component of the system. So, from a systems point of view, it's helpful to consider these outputs in a stocks-and-flows framework.

Flows refers to the things that are being produced by the system (grain, fruit, fish, fiber, timber, and so forth). *Stocks* refers to the components of the system that produce these flows, or the components that have value in their own right. Stocks, therefore, include things like healthy soil (that grows the grain) or the fish stock from which the harvested fish are caught (and which, in turn, depends on stocks of plankton they feed on) or the biodiversity that people want.

Stocks are the foundation of our social, economic, and ecological activity. They are the source of our prosperity and "wealth." Resilience practice is really about understanding and managing the resilience of these stocks. Flows are valuable in an immediate sense as a means of identifying the underpinning stocks—which are really the system's assets. Focusing only on the flows and basing your management on maintaining them without consideration of the stocks can have catastrophic consequences. Consider the example of the Aral Sea in box 3 (page 46).

Keep in mind that the same stocks can be used to provide different flows. And, in some cases, flows and stocks can get a bit confusing. Consider biodiversity. It can be a stock that underpins the regenerative capacity of soils or rivers or lakes, and it can also be a kind of flow when it

---

Box 2: An Ecosystem Services Lens

From 2001 to 2005 the Millennium Ecosystem Assessment sought to appraise the condition and trends in the world's ecosystems and the services they provide. It grouped ecosystem services into four categories:

- Provisioning services: These give us things we use and need directly, including food (crops, seafood, etc.), water, pharmaceuticals and other products of that kind, and energy (hydropower, biomass fuels).
- Regulating services: These include pollination of crops, pest and disease control, processing of wastes, water and air purification, and carbon sequestration.
- Supporting services: Soil regeneration, nutrient cycling, and seed dispersal are some ways ecosystems support us.
- Cultural services: Examples include recreation (including tourism), intellectual stimulation, customary practices, and spiritual inspiration.

These are general classes, and the first step is to identify within each class the main services that are important in your system. Next, think about how these services are connected—the idea of ecosystem "bundles"—groups of services that tightly interact and involve trade-offs between each other.

An example of trade-offs between ecosystem services can be seen in the Camargue wetlands in France. Depending on how they are managed, these wetlands might exist as reedbeds that provide reeds for the traditional thatchers and act as fish hatcheries; or as open water that provides habitat for ducks, much valued by hunters; or as wet meadows

Continued on page 44

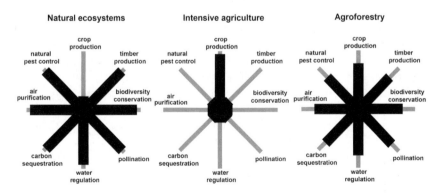

Figure 5: Variations in Bundles of Different Ecosystem Services with Variations in Land Use

(After Foley et al. 2005.)

Box 2 continued from page 43

that provide good pastures for the famed Camargue horses. Manipulating the Camargue delta to increase one service can result in the reduction of another. Hence the term bundle of services.

Figure 5 is a general depiction of how eight ecosystem services are provided to differing extents under three different forms of land use. Deriving some version of this figure for different management options of your system is one way to begin addressing the trade-offs involved in choosing between the different options.

Two questions immediately arise: How do the services interact? Are there any likely threshold effects? That is, if you increase the flow of one particular service, what is the likelihood that, as a consequence, another one may cross some critical threshold and not be able to be restored? Such an effect could significantly change the estimates of economic benefits from different management options.

Raudsepp-Hearne and colleagues (2010) provide an actual example of the trade-offs involved among twelve ecosystem services in six kinds of social-ecological systems in 137 municipalities in Quebec. They grouped the twelve services into provisioning (crops, pork, drinking water, maple syrup), cultural (deer hunting, tourism, nature appreciation, summer cottages, forest recreation), and regulating (carbon sequestration, soil phosphorus retention, soil organic matter). They analyzed the trade-offs among the twelve services.

The main trade-off was between provisioning services and both regulating and cultural services. Crops and pork production (provisioning services) had the highest negative impacts on the other services. There were also synergies among services. All regulating ecosystem services were positively correlated with each other. Raudsepp-Hearne and colleagues noted in particular that soil phosphorus retention had a strong positive correlation with the quality of drinking water. Both were negatively correlated with pork production, a large emitter of phosphorus.

Three verifiable critical threshold levels were found in the services—soil phosphorus (P) retention (the threshold is where P saturation > 12 percent), soil organic matter (3.4 percent), and drinking water quality (Quebec Water Quality Indicator < 3.5). Beyond any of these thresholds there are negative consequences for human well-being. With the help of a spatial mapping exercise, Raudsepp-Hearne and colleagues showed that fifty municipalities had crossed at least one threshold, five had crossed two, and four had crossed all three thresholds—all in the feedlot agriculture bundle type.

The increasing recognition of the social, ecological, and economic importance of unmarketed ecosystem services places growing attention

Continued on page 45

on the trade-offs involved. These trade-offs are especially important, but particularly difficult to overcome, when it comes to common-property issues. In fact, the recognition of ecosystem services sometimes results in a change from what was previously regarded as a private-property to a common-property asset. Obvious examples are water and air pollution resulting from intensive agriculture.

There are two important take-away messages from this. The first is that an ecosystem services approach is a basis for a level playing field across sectors. Each sector will have different economic values for the different flows in terms of output or people employed, or valued stocks (like biodiversity), and it's difficult to compare the net worth or "value" of the sectors. However, using an ecosystem services approach it is possible to show how the impact on an ecosystem service in one sector affects the flows that are valued by other sectors. And understanding interactions between ecosystem services (considering them as linked bundles) makes it possible to consider the secondary effects of changes in the system.

The second message is that from a resilience perspective, the most important thing to consider is the nature of the interactions (trade-offs) between services. Are they all smooth, or are there sudden, marked changes or thresholds? If a change in one ecosystem service results in a precipitous, and perhaps irreversible, change in another, it's important to know about it.

---

is regarded as something of value in itself. This is not a problem. You just need to be clear about how you are dealing with it in each case.

The aim of this part of the assessment is to identify, and get agreement on, the important system goods and services and the stocks that underpin them in the system, and then to examine the interactions between them. (Note that we say "system" and not "ecosystem" goods and services. Sometimes people put high value on heritage sites and non-ecosystem-based livelihoods as valued goods and services. Their resilience is also important).

The Millennium Ecosystem Assessment uses the phrase *bundles of ecosystem services* to capture the fact that groups of ecosystem services interact; if one goes up, others may go down, or also up. They are not independent of each other, and very often there are trade-offs between them. If the shapes of these trade-offs have sharp changes, or thresholds, it is important to know about them, for they may indicate critical thresholds in the resilience of the delivery of ecosystem goods and services.

In your system, what trade-offs are occurring among the valued

Box 3: Water Stock and Cotton Flow–The Aral Sea Saga

Ignoring stocks and the factors that underpin their health can result in their loss. Possibly the best example of this is what happened with irrigation around the Aral Sea, where the system's focus was completely on the flows without any consideration of the stocks that provided them.

The Aral Sea began drying out in the 1960s when irrigated cotton production was dramatically expanded under a Soviet command-and-control enterprise. The stock was the available water, and the flow was cotton grown on irrigated land. The expansion of irrigated agriculture resulted in soil degradation, salinization, and the shrinkage of the Aral Sea. The response to these declines was only to take action that sustained cotton yields. Little attention was paid to the declining stock that enabled cotton production. The response, therefore, was the expansion of the amount of land under irrigation and the implementation of technical interventions such as large drainage schemes, which allowed the application of more water to leach salt out of the surface layers prior to irrigation.

The region entered into a positive feedback loop of environmental degradation, which prompted further expansion, which in turn caused further degradation. The cycle was maintained by demands from Moscow (control from a higher scale) for cotton deliveries, focusing solely on yields and disregarding water use (Schluter and Herrfahrdt-Pahle 2011).

In the short term, this ensured a stable flow of cotton and economic rents, thereby creating vested interests and rent seeking that perpetuated the cycle. In the long run, it has seen the stocks that provided this flow fall into irreversible decline. Indeed, the process has devastated the entire Aral Sea basin (something that is discussed in greater detail in chapter 4). The region remains in this vicious cycle today.

While this is an extreme example, it's difficult to imagine that planners in the region couldn't see what was happening. Indeed, in interviews carried out by Schluter and Herrfahrdt-Pahle, one planner acknowledged anticipation of a decline but believed the situation to be solvable through the application of technical measures such as getting water from elsewhere, in this case diversion from Siberian rivers. While this was actively contemplated in the dying days of the Soviet Empire, it was never implemented because of the prohibitive cost. And the scale of the problem has defeated all successive efforts.

system services? Are there examples of private property assets that are in fact functioning as common property, and are there any resilience issues involved? What are the shapes of the relationships between the pairs of ecosystem services—do any of them exhibit sharp changes or threshold-like effects? Does this call for a change in the way the ecosystem services are managed, and regulated? This comes back to the issue of governance and whether the current set of institutions is appropriate for the management of common property. Governance is discussed further in chapter 4.

The Big Issues

It is useful somewhere around here to ask, What are the big issues? What are people worried about? Often the issues are concerns about the viability of the things people value, and so the same things come up. Moving between "values" and "issues" can be a useful way to proceed in identifying the "resilience of what." Sometimes the big issue may seem purely social, not directly related to natural resources. It's important to be open to discussing any issue raised by stakeholders. An issue might sound trivial to begin with but turn out to be significant. In one agricultural region, for example, the problem of daughters-in-law was raised. At first glance it seemed quite trivial, but as it became apparent that it was a widespread problem, its significance grew.

The problem of daughters-in-law was that most of the farms in this region were being run by elderly men who had not yet handed their farms over to their sons due to concern that divorce settlements were awarding half of each farm to each partner, regardless of family history. Several of the daughters-in-law, disaffected because they felt taken for granted and forgotten, had responded by leaving. They took their children with them. Often the sons went as well. At the scale of the farm, this can be disastrous, as it prevents revitalization with new blood and complicates intergenerational transitions. But if the problem is widespread enough (and at one particular resilience workshop it clearly was, since many people raised the issue), then the impact will also be felt at higher scales as rural towns are stultified by a lack of new people and the skills they bring. It reduces the capacity of farming regions to adapt and reduces their openness to novel ideas and new ways of doing things.

This is an issue that sits in the social domain of a social-ecological system, but it clearly influences the general resilience of the combined system. It also links to higher scales.

A social issue identified in the Goulburn-Broken catchment in Australia (discussed in chapter 3) was the growing awareness that more of the available water should be reserved for environmental flows. This was a shift away from the more traditional belief that water should go to uses that maximize its economic value. But the people who believed the environment should get more of the available water were largely based in the cities. In the region, at the focal scale, there was considerable resistance to this (though it is changing over time).

By the end of this step, you will have developed the set of the things considered to be important in and about your system, and the stakeholders' concerns about them. You may have also assigned each item in this set to a scale and have notes on how issues connect across scales and domains. Importantly, you will have (or should have) defined the important assets in the system that underpin the valued system goods and services and gotten an initial idea of the interactions among them.

These are the things about the system that you want to make resilient. You don't want them lost or diminished by some act of God or through ignorant mismanagement. These are the things that are worth spending time and effort on, the things that are the answers to the questions, What do you want to make resilient? and The resilience of what?

4. The Resilience to What? (Disturbances to the System)

So far, we have made an effort to map the system over the scales at which it functions; we've identified the main players and asked what's important about the system (the resilience of what). Now we begin to attempt to characterize the dynamic nature of the system by looking at what the system has to deal with in terms of disturbances. It is useful to do this for three categories: characteristic disturbances, large infrequent disturbances, and unknown shocks.

Characteristic Disturbances

These are the disturbances you know and expect. They might be related to monsoonal flooding in many tropical areas, or bushfires or droughts in drier regions, or severe frosts in temperate zones. They may occur every year or strike with sufficient frequency that you (and the ecosystems) aren't "surprised" when they do.

They are disturbances under which the system has evolved. It is adapted to these disturbances and deals with them quite well; indeed, it's by dealing with them that it is resilient. A flood or drought might temporarily hurt the things that you want out of your system (like agricultural production), but while doing so it also gives a boost to the species/varieties that can deal with a flood or a drought, and the system returns to "normal" with relative ease following such a characteristic disturbance.

As discussed in chapter 1, resilience practice is not about stopping change or eliminating disturbance. Efforts to prevent characteristic disturbances, such as prescribed burning to stop wildfires or damming to control floods and prevent droughts, may generate desirable outcomes over short time frames but inevitably lead to larger, more intractable problems over longer time frames. In fact, the real "disturbance" to the system in terms of these kinds of characteristic disturbances is the change to the characteristic disturbance regime. That can lead to loss of resilience.

Large Infrequent Disturbances

These disturbances are often similar in type to characteristic disturbances but are rarer and significantly larger in magnitude. The once-in-a-century flood, the every-so-often outbreak of a known pest, the wildfire that breaks all the records, and the large earthquake that hits the city dead center are all examples of large infrequent events. These are events that may have happened in history but that don't seem likely in the near future. The system has not had sufficient experience with these kinds of disturbance to have evolved mechanisms for dealing with them.

These are events that can reconfigure a region, pushing the system into an alternative regime. Sometimes that alternative regime may be undesirable because "business as usual" is no longer possible, but sometimes these disturbances constitute renewal events. Hurricane Katrina reconfigured New Orleans. Large storms are common to that part of the world, but the magnitude of Hurricane Katrina swamped the city's capacity to absorb the disturbance.

Unknown Shocks

These are disturbances that come out of left field. You cannot predict them, and it's almost impossible to prepare for them. An unknown shock might relate to an invasion by a previously unknown exotic pest or weed or to the outbreak of previously unknown disease. It might be a terrorist attack in a place that has never seen terrorism, or a tsunami where none had ever been experienced. It might be an emerging technology that disrupts existing enterprises while creating new ones.

Unknown shocks are sometimes referred to as "black swan events." In the Northern Hemisphere it was thought that there were only white swans until black swans were discovered in Australia. Prior to the discovery of black swans by Europeans, the term was used to describe something that couldn't exist. However, after it was discovered that black swans did exist, the term *black swans* changed in meaning to describe events that might exist but which we cannot anticipate.

Work is currently under way by researchers at the Stockholm Resilience Centre to draw up an "architecture of surprise" in which three archetypal forms of unexpected crisis are described (Homer-Dixon et al. In prep). The first is the "multiple whammy," in which simultaneous stresses within a single system synergistically combine to produce overload. Then there's the "long fuse, big bang," arising from the slow accumulation of one or more stresses ultimately producing a sharp nonlinear shift in the system's behavior, perhaps toward a new stability regime. And the third archetype is the "ramifying cascade," arising when a sudden and severe perturbation in a tightly linked network propagates through the network, producing a range of significant secondary effects. This analysis should be available in the near future.

Attempting to list the disturbances your system experiences is part of describing the "resilience to what." What is it that you want your system to be able to recover from? A good way of exploring this further is to identify the kinds of things that have caused big changes in the past, and a good way to do this is by reviewing the history of your system in terms of the drivers that sculpted it.

5. Drivers and Trends (History and Futures)

Your system is unique. There's nothing exactly like it anywhere. How did it get to be that way? What were, and what now are, the "drivers" of system change? Exploring the history of your system—developing a time line—is likely to provide interesting insights on these questions.

Get the stakeholders to develop a historical profile of their system. One way of investigating this is to use a whiteboard (or even the increasingly rare blackboard!) and draw a time line across it, beginning from pre-"development" days (usually one to three hundred years ago). Better still, set up three time lines: one for your focal scale and one on either side for the scale below it and above it. Identify things that caused changes in the system. In one workshop, people chose to use three rolls of paper across the floor of a room.

Mark specific events and changes onto the time line. Things like major climatic events, introduction of new technologies, new crops, war- and postwar-related changes (like soldier settlement schemes), major infrastructure developments (railways, dams), introduction of significant legislation, appearance of new pests or diseases that caused big changes, and so on.

To outsiders this step may appear a bit boring. Who cares that the Great Storm occurred in 1948. But for stakeholders this is usually a stimulating exercise as they explore their common understanding of why their system (farm, catchment, region, and so forth) is the way that it is. Its great value is that it generates insights into event-driven changes, cause and effect, and what's really important in the system. You're almost guaranteed to discover things that will reframe other aspects of your system's description.

The use of three time lines (for three scales) can provide interesting insights into cross-scale causality, where a significant event at one scale matches with (is caused or followed by) a significant but different change at another scale. Such a history of your system is a good starting point for considering current trends that are causing change and for building scenarios of where the system might be heading.

What are the current drivers at each scale? What trends are occurring at national and global scales? Remember, drivers are external to the system (or from a higher scale)—we take them as inputs from outside and don't consider feedbacks from our system to them, even though they may be occurring. For example, climate change is a driver of change at an ecosystem or farm or forest scale, but even though what happens in those systems does have some feedback to the climate, we do not consider that in our analysis of resilience.

This is part of bounding the problem to develop the system we consider self-organizing. It is important to do this because in the next stage of the practice (assessing resilience; see chapter 3) we'll be identifying controlling variables in the system, and there is sometimes confusion about the distinction between drivers and controlling variables. So,

Table 1. Drivers, Variables, and Thresholds

System	Driver	Controlling variable	Threshold
Agricultural regions	Clearing of native vegetation (due to rising demand for agricultural production)	Depth of groundwater	Two meters below surface
Coral reefs	Use of agricultural fertilizers on land	Nutrient level in water	Nutrient level above which algal growth removes space for coral polyps
Coral reefs	Fishing pressure (due to population pressure and demand for fish)	Herbivory	Herbivory level below which algal growth preempts space for coral polyps
Tropical forestry	Agricultural intensification (due to rising demand for forest products)	Interval between "fallows" and reforestation	Level of nutrient depletion at which tree seeds don't germinate

let's take a moment to clarify the difference. Drivers cause changes in the controlling variables. Consider table 1. It provides five examples of drivers and controlling variables.

Notice that table 1 is scale dependent. For example, at scales larger than agricultural regions (i.e., countries, the world), the driver is population growth or economic demand, which causes changes in the controlling variable, native vegetation cover, which has a threshold effect that causes water tables to rise rather than fall. Once again, you need to define the focal scale, because all complex system dynamics are scale dependent.

Identifying trends that may be acting as drivers of change amounts to asking what trajectory (or trajectories) your system is on. Is it getting drier or wetter or hotter? Are farms getting bigger? Is your region experiencing changes in demography (age distributions and population size)?

Joining the Dots

And that, ladies and gentlemen, is your system. It's not a perfect description, but it's a good starting point for engaging with its inherent complexity. You might also like to think of it as bringing together the various pieces that make up the jigsaw puzzle that *is* your system. Now the challenge is to assemble these pieces to build a picture of how the system is changing and to consider how much it can change before it can no longer recover (before it loses or changes its identity).

Key Points for Resilience Practice

- A good description brings together the insights of the key stakeholders in the system.
- Good resilience practice is not so much about producing a single "best" system description as it is about creating a process whereby the system description is constantly revisited, reiterated, and fed into adaptive management.
- It's about the balance between including all the critical information and requisite simplicity (as simple as possible, but not too simple).
- Understanding what's important in your system in terms of valued goods and services and the stocks that underpin them, and the interactions among these bundles of system goods and services, is a good way to begin coming to grips with the "resilience of what."
- Drivers cause change in controlling variables; as a controlling variable approaches a threshold level, a shock to a fast variable (goods/services), or a directional change in a driver, can push a system across a threshold into an alternate stability regime.

CASE STUDY 2

From Taos to Bali and Sri Lanka:

Traditional Irrigation at the Crossroads

At first glance the acequia farmers in New Mexico and the *subak* rice growers of Bali don't have much in common beyond the fact that they both use irrigation in their farming enterprise (see images 3 and 4). They produce different crops in dramatically different environments using different traditions. However, dig a little deeper and it's clear there are many similarities underlying their resilience, a property that has seen them weather many disturbances and yet continue to operate successfully over centuries. Also in common is a growing vulnerability to modern growth and globalization.

The Acequias of the Taos Valley in New Mexico

In the Taos Valley in northern New Mexico, farmers have been successfully growing crops for centuries despite frequent droughts and low annual rainfall (just over three hundred millimeters a year). They farm using irrigation water provided by acequias. *Acequia* is a Spanish word adopted from Arabic, describing both an irrigation canal or ditch and a community of irrigators. The water the Taos Valley acequias use is largely supplied by snowmelt from the Sangre de Cristo Mountains that border the valley to the east and southeast. We are indebted to Michael Cox at Indiana University for the details of this story.

The acequias farmers are the descendants of the Spanish colonists who moved north along the Rio Grande from Mexico around 1600, bringing with them Spanish irrigation traditions. One important in-

Image 3

An acequia in the Taos Valley. (Photo: M. Cox.)

stitutional regime underpinning their practice is the notion of water being a common property within each community acequia. Compliance with community obligations is required in order for any person to maintain individual water rights.

Acequia farmers use a form of flood irrigation involving the clearing of unlined ditches in the ground in order to convey water to their fields. Each acequia community employs a mix of private and common property rights. These are implemented mostly by three elected commissioners and an executive majordomo. Each acequia member owns a private parcel of land, as well as the ditch that immediately feeds it. The larger community ditch and the water running through it are common property, meaning that to access water each member must observe an established set of community rules.

Water and land rights within acequias are largely independent of the state government. The majordomo is in charge of allocating water within each acequia, and the commissioners serve a variety of legal roles. The commissioners are frequently called on to rule on disputes and sup-

Image 4

Irrigation diversion channels in Balinese rice fields (with a small water temple). (Photo: S. Lansing.)

port the majordomo in enforcing "ditch rules." Their "peon" system of proportioning costs and benefits maintains a sense of fairness among members with unequal wealth, and their rules around water distribution help maintain the decentralized and low-cost monitoring of infractions.

When disputes arise between acequias, no single acequia has authority over the others. Instead, when droughts occur majordomos and/or commissioners from a set of acequias may meet in accordance with historical water-sharing agreements. These agreements play an important role in the acequias' responses to drought by ensuring that available water is shared in a way acceptable to the community.

The acequia communities form a modular network in which there is much more intensive and regular interaction among farmers within acequias than between acequias. These interactions revolve around water distribution, monitoring, and the resolution of conflicts. Weaker connections exist between many acequias. These links are still important, as they enable the communities to resolve collective-action problems on a larger scale through water-sharing agreements.

The modularity of the overall network accomplishes several things. Primarily, it breaks the larger irrigation system down into subgroups, and each group can resolve collective-action problems independently of other groups. This decreases the number of individuals, and thus the transaction costs, involved in resolving any particular collective-action problem.

And the system has proved very resilient. The acequias' sustainability in the face of severe droughts has enabled them to persist as community-based irrigation systems for several hundred years in a high desert environment. But the farming communities are now facing threats that did not play a part in their evolution—economic development, changing demographics, and the penetration of water markets.

New users are settling within the territories historically irrigated by the acequias, and this has led to the fragmentation of property rights, as those rights must be subdivided among a larger number of users. This leads to increases in transaction costs and makes the maintenance of collective action more difficult.

Newcomers also increase the cultural difference in the affected acequias, making it harder to maintain traditions. New settlers can be highly disruptive when they generate higher levels of conflict over the interpretation and application of rules. This substantially increases enforcement costs.

And with more people there has been a growth in urbanization, with the creation of new roads and parking lots and the addition of many impervious surfaces that change the area's hydrology. Further interfering with hydrological flows to acequias' canals, the dense population within the town of Taos, the urban center of Taos Valley, makes high local demands on water. This makes collective action more difficult close to the town, as it can disrupt the flow of benefits that participants might obtain through cooperation.

An analysis of the Taos Valley acequias by Cox and Ross (2011) found that the ability to maintain collective action is aided by water-sharing agreements and access to groundwater. However, collective action is hampered by the fragmentation of property rights and urbanization. They found that smaller acequias are better able to maintain crop production per unit area than larger ones.

If stakeholders in this region were to attempt to describe their system as part of a resilience assessment, it would be important to incorporate these different aspects of modular structure, scale, governance, trends, and disturbance regimes. And by *disturbance*s we mean those that the system is traditionally accustomed to cope with as well as novel disturbances.

In many areas around the world, the increasing economic connectivity is the source of novel disturbances that impinge on the historical practices of community-based management that employ common-property arrangements. Such connectivity may afford new opportunities, but it also comes with costs. In this case, Cox and Ross (2011) found the cost has involved decreased interdependence and solidarity both within and between acequias, as they are less involved in historical traditions and rituals and more involved in the larger and more developed economy through wage-earning jobs and local markets. Cox and Ross found that the acequias that have been more exposed to the novel disturbances of land-rights fragmentation and urbanization are producing fewer crops per unit area than other acequias in the valley.

The Bali Water Temples and Rice System

With its combination of tropical climate, high rainfall, and fertile volcanic soil, Bali couldn't be more different from the Taos Valley. Over the last thousand years the people of Bali have extensively modified the landscape of their island, terracing hillsides and digging canals to irrigate the land in order to grow rice. And yet the irrigation traditions practiced in Bali exhibit many similarities to the acequias traditions of Taos. The following account is based on the work and experience of Steve Lansing and Rachel and Stephan Lorenzen, who have spent many years studying the culture underpinning Bali rice farming.

Rivers run all over Bali, and rice farmers have formed themselves into cooperatives known as *subaks* to develop an elaborate irrigation infrastructure. The *subak* is a unique social and religious institution, a self-governing organization of farmers who share responsibility for the just and efficient use of irrigation water to grow paddy rice. Farmers whose fields are fed by the same water source meet regularly to coordinate plantings (important for avoiding pest damage), control the distribution of irrigation water, and plan the construction and maintenance of canals and dams. They also organize ritual offerings and *subak* temple festivals. All *subaks* possess legal codes, detailing the rights and responsibilities of membership.

Balinese rice paddies are built around the bases and slopes of the island's volcanoes. The flanks of the volcanoes are cut by ravines containing small rivers and streams. Over the centuries the farmers have constructed hundreds of small diversionary dams and weirs at the bottom of these ravines. Each weir diverts the flow from a short stretch of river into a small irrigation channel or tunnel. These angle sideways and emerge nearby or more than a kilometer downslope to flood blocks of rice terraces that have been carved into the flanks of the volcanoes. Each *subak* manages several blocks of terraces. Water flows in huge quantities in the rainy season, less in the dry season. Using traditional engineering techniques, the *subak* is able to provide water equally to each unit and subunit, as many *subaks* are subdivided.

In addition to the engineered landscape of dams, tunnels, aqueducts, and rice terraces, each *subak* maintains a local network of shrines and water temples, where farmers make offerings to their gods, and there are also temples on the inter-*subak* level where water is divided to serve several *subaks*. As well as honoring the gods, the interlocking cycles of water temple rites also provide a template for the management of ecological processes in the paddy fields. By adjusting the flow of irrigation water, farmers can exert control over many ecological processes in their fields. They coordinate cultivation cycles, so water to one *subak* may be diminished while another *subak* is beginning its planting cycle. Mineral nutrients needed by the rice are leached from the volcanic landscape by the monsoon rains and transported to the fields in the irrigation water.

Water can also be used to manage rice pests (rodents, insects, and diseases) by synchronizing fallow periods in large contiguous blocks of rice terraces, depriving pests of their habitat and thus causing their numbers to dwindle. Pest reduction is also achieved through keeping ducks in the field before the plowing season and flooding. The success of this method of pest control depends on the ability of the farmers to schedule simultaneous harvests over quite large areas so that the pests cannot migrate to a new food source. Simultaneous cropping is done *subak* by *subak* (or sometimes in subunits), which requires a smoothly functioning, cooperative system of water management.

Lansing believes that the resilience of the *subaks* is due to the water temple networks because they enable a bottom-up coordination that is essential for effective production (Lansing and Fox 2011). Controlling pests, for example, involves a fine balance. To reduce losses to pests by fallowing, farmers must take into account the population dynamics of the most damaging local rice pests: how fast can they reproduce and migrate, and how much damage can they cause. Synchronized fallowing has implications for water sharing, because peak irrigation demand in rice paddies occurs at the beginning of the cultivation sequence. Water sharing and pest control by synchronous harvests are thus opposing constraints.

Consequently, irrigation schedules in a Balinese watershed are like puzzle pieces. If they fit together well, the resulting pattern optimizes harvests for everyone by minimizing pest infestations and reducing water shortages. But if too many fields go fallow at the same time, there will be water shortages downstream when they are all reflooded. Conversely, if the areas of synchronized fallow are too small, pests can migrate to adjacent fields that are still in cultivation. Water temple networks enable the farmers to adjust their irrigation schedules in response to local conditions, solving the jigsaw puzzle for entire watersheds that may include dozens of weirs and local irrigation systems.

Multi-*subak* groupings form the congregations of regional water temples. They exist to acknowledge the sites where water originates, such as crater lakes and springs, and where the water is divided. All of the farmers who benefit from a particular flow of water share an obligation to provide offerings at the temple where their water originates. For example, if six *subaks* obtain water from a given source, all six belong to the congregation of the water temple associated with that source. Thus the larger the water source, the larger the congregation of the water temple.

It's a system that has served the Balinese farmers well over many centuries. Communal *subak* work obligations substantially reduce labor requirements of individual households and support Bali's particular methods of rice cultivation, which are seen as the foundation of regular high yields. Synchronization of irrigation and ceremonies means fewer pests, joint operation and maintenance of the irrigation system mean less work for the individual, and joint water management means group pressure on free riders (Lorenzen and Lorenzen 2010).

But, as with the farmers of the Taos Valley, more people and greater connection with a globalized world are presenting a new set of challenges for the *subak* tradition. Golf courses and new tourism developments are increasing the demand for water. The biggest water consumer in Bali is the hospitality industry, and demand is driving up its price.

In 2010 there were approximately 82,000 hectares of irrigated rice terraces in Bali. During the period 2002–2006 an average of 641 hectares per year were converted to other uses. This rate of loss is estimated by the Agriculture Ministry in Bali to have increased to about 1,000 hectares per year.

Although rice farming continues, for many households it has become a side business, with tourism and employment in the urban centers providing alternate sources of income. Historically strong ties through the *subak* with other elements of Balinese culture, such as the village and the temple congregation, are clearly weakening because rice cultivation is no longer the major economic activity of most Balinese. And with rice fields continuing to disappear and fewer incentives for the younger generations to engage in agriculture, the *subak* as a form of Balinese social organization and way of cooperation is endangered.

Part of the change has been a shift from communal to outsourced labor for the peak labor periods in the cultivation cycle (Lorenzen and Lorenzen 2010). The disappearance of communal work arrangements is a clear sign of a shift from subsistence-oriented agriculture to a more individualized and market-oriented system. Nevertheless, this move is regarded as positive by both older and younger farming household members because they associate the monetizing of irrigated agriculture with greater access to modern consumer goods.

The older generation still remembers much harder work in the rice fields in the past, periods of starvation between harvests, and the overall higher poverty rate within the villages. The farming community is aging, and better education and off-farm employment opportunities are increasingly available to the younger generation (Lorenzen and Lorenzen 2010).

Choices, however, depend on individual household circumstances, such as land available for sharecropping as well as the skills to pursue off-farm nonagricultural work. And the rice field still holds some importance as it offers a measure of social security in an insecure world. In the wake of the Bali bombings and the global economic crisis, for example, tourist numbers significantly dropped, resulting in fewer permanent and casual off-farm work opportunities. Those who had kept their rice fields while working off-farm considered growing rice again, while families who had opted to sell their fields ended up with serious financial difficulties (Lorenzen and Lorenzen 2010).

In addition to providing a degree of financial security, the rice terraces are precious for their iconic value. They are a major selling point in the Balinese tourist industry. However, if they are not to disappear, incentives have to be created to keep them viable.

Young people are more interested in global developments, moving to cities, and not getting their hands "muddy" as their parents do. The famous Balinese rice terraces are under growing threat from the pull from higher scales in the form of new (globalized) jobs. And this has a negative impact on human and social capital at the focal scale—not just a shortage of labor but the inheritance of knowledge and trust that allows the necessary sharing and cooperation to occur.

Other cultures where traditional irrigation schemes have been lost are seeking to reinstate them. One example is in Sri Lanka.

The Sri Lankan Tank System

An interesting contrast to Taos and Bali comes from an even more ancient system—the Sri Lankan "tank" irrigation system. It was developed in the drier, monsoonal northwest of Sri Lanka over a period of sixteen hundred years—from the third century BC to around 1200 AD. It was very large and complex, with around twelve thousand tanks supporting some twenty thousand small irrigation systems, each less than eighty hectares, via a myriad of irrigation channels.

As described by Panabokke and colleagues (2002), it is a cascading network in which water flows from one tank, along a channel with paddies on either side, into the next one, and so on. The tanks were apparently originally developed more for household and livestock use, with irrigation starting later. The traditionally self-sufficient lifestyle was based on rain-fed *chena* cultivation, lowland tank-irrigated rice cultivation, homestead mixed-garden farming, cattle grazing and herding, tank fish harvesting, and food gathering.

After 1200 AD, with a change in ruler, there was a progressive decline in the system until only around half the original tanks remained. But they were still very important, socially as well as for irrigation. The traditional method of tank maintenance was the *Rajakariya* system, which drew upon the collective effort of the farmers under the direction of a *Velvidane* to implement decisions on water distribution and use made by the community in *Kanna* meetings. It was quite a complex system, involving discrete communities linked by systems of caste and kinship, sharing access to a range of common-property resources, including the tanks.

The British abolished the *Rajakariya* in 1832, as it was considered a form of slavery, and the tanks fell into further disrepair. But many still persisted and continued to meet societal needs. In the 1950s the resources were effectively nationalized by the state, which led to further deterioration. The system then evolved toward larger dams, and recently the emphasis has been on a few much larger, more cost-efficient dams. Most recently, like Bali and Taos, the effects of new markets and economic pressures are leading to further changes.

There was little documentation of just how the original tanks were managed, but the cultural aspects persisted and are now assuming increasing importance. Like the Bali system, the irrigation systems were associated with temples, and the social system was based on a triangular representation of tanks-rice-temples, each being intimately connected to and dependent on the other two.

What is interesting and different about this system is that most recently there has been a cultural revival of interest in maintaining the tanks and temples. Countering the threats by globalization and modern development pressures, the ancient songs associated with the tanks and their temples are being revived and celebrated, and the societies involved want the systems restored and maintained (Srikantha Herath, personal communication).

Whereas in the Taos Valley and Bali customary approaches are being eroded by outside and higher-scale forces, in Sri Lanka there is hope of reviving old ways to better cope with new times. It's believed this will help build social capital, cooperation, and trust.

Resilience Practice and Traditional Irrigation (When the New Displaces the Old)

The lesson for resilience practice in these stories of irrigated farming is that to understand the resilience of these systems, it's important to appreciate their modular structure, governance, and scale and the nature of cross-scale impacts.

In both Taos and Bali, the irrigation tradition is based on a modular network of user groups who make decisions within groups and/or between groups as circumstances demand. Water and the infrastructure to deliver it are seen as a common property that all users have rights and responsibilities for.

While all the traditions have served farmers well over several hundred years, they are not insulating them from disturbances associated with population growth, urbanization, and globalization.

So another lesson is that sometimes older, evolved systems become almost maladaptive in the face of higher-scale developments. When this happens transformative change may be inevitable. If so, the choice is then between doing nothing, and having to endure the social and other costs of whatever the transformation process brings, or actively guiding the transformation process so as to retain the valued aspects of the "old" system while achieving a system that is in tune with the larger world.

3

Assessing Resilience

H aving developed an agreed-on description of the system, the next step is to assess its resilience. The process here is not feeding your description into some formula. Resilience is not a single number or a result. It's an emergent property that applies in different ways and in the different domains that make up your system. It is contextual and it depends on which part of the system you're looking at and what questions you're asking.

So the next step is arranging the components you've described into an order that gives you some insight into how your system is behaving and changing—its dynamics over time. And, given those dynamics, what are the things you need to be most careful about?

The task of assessing resilience encompasses understanding both specified resilience and general resilience. As discussed in chapter 1, they are complementary, linked aspects of self-organizing social-ecological systems. But a full assessment also entails understanding the system's capacity for transformational change (transformability)— what capacities a system needs in order to reinvent itself. This chapter, therefore, is divided into three sections describing these qualities.

We're presenting them in this order (specified resilience, then general resilience, and finally transformability) because it has proved to be useful in a number of case studies. However, we note that some groups have preferred to begin with general resilience and then to come down to thresholds as particular aspects of resilience. In the spirit of the adaptive approach that is basic to resilience thinking, we applaud the diversity of approaches. Whatever works best for you is the way to go. Don't get bogged down in trying to complete one aspect of the assessment in its entirety. The most important thing is to iterate between them.

Specified resilience, general resilience, and transformability lie at the heart of resilience practice. Understand these concepts and everything else follows.

Specified Resilience

Specified resilience is the resilience of some part of the system to particular kinds of disturbance. Of most importance, it's about whether a disturbance might push the system over a particular threshold where it changes the way it functions (e.g., stops producing grain or timber or providing habitat). The aim here is to identify known and possible thresholds between alternate states (or regimes) the system can be in.

Thresholds occur on underlying, controlling variables, which often change slowly relative to the variables you're concerned about. For example, the variable of concern might be crop production and the controlling variable along which a threshold for crop production lies might be soil acidity. Indeed, this soil acidity threshold was identified as an important threshold for the Goulburn-Broken catchment in Australia (see figure 7, page 71).

Because controlling variables often change slowly, the changes tend not to get noticed by managers, and so thresholds are often not factored in. Take another look at figure 1 and the discussion on thresholds in chapter 1 to remind yourself how different types of thresholds behave.

Depending on the kind of shock experienced by the system, some thresholds are more or less likely to be crossed than others, and the subsequent trajectory of the system depends on which one gets crossed first. It can lead to a cascading effect of others being crossed.

The aim in attempting to assess specified resilience is to produce some form of representation of the system that shows possible thresholds and how they might interact with each other. Because self-organizing systems operate at different scales and in all three domains—social, economic, and ecological (biophysical)—thresholds can occur in each domain and at each scale.

For this discussion let's use a basic situation—a focal scale with one scale above and one scale below. To begin with, therefore, the framework you're attempting to create might depict the thresholds in nine categories ("boxes") in a 3 scales x 3 domains matrix (as first described in Kinzig et al. 2006). We've illustrated this in figure 6a. In this representation the three scales being considered are the patch, the farm, and the region. Our 3x3 matrix has double-pointed arrows between each threshold box, suggesting each threshold interacts with the thresholds around it in different domains and at different scales.

Keep in mind this is only an idealized version of a thresholds matrix. It's unlikely that each slot will have a threshold, and it's possible some boxes will contain more than one. At the farm scale, for example, there might be a biophysical threshold relating to the depth of groundwater and another relating to a level of soil acidity. Cross either threshold and the farm might tip into an alternate state (of low productivity and subsequent bankruptcy). And these thresholds might be independent of each other.

Figure 6b represents a thresholds matrix for dry forest systems in a region of Madagascar (Kinzig et al. 2006 and Bodin et al. 2006). It can be readily compared to the idealized 3x3 matrix, but there aren't thresholds in each box. The scientists who were working on this system didn't believe there were any threshold effects operating in several of the slots (at the farm/biophysical level, for example).

They also proposed how they believe these various thresholds interact with each other. For example, at the regional level the threshold of concern related to further loss of forest and increasing fragmentation of what was left. Beyond a certain level of fragmentation, they believed, there will be a loss of pollination as an ecosystem service at the patch scale because the pollinators live in forests. There was evidence of declining levels of pollination with increasing distance from the edge of a forest. This will have an impact on crop production levels at the farm scale and could lead to some farms crossing an economic viability threshold—a cascading effect.

Let us now consider a more complex thresholds matrix in which several thresholds are believed to be operating in a single domain/ scale. The thresholds matrix presented in figure 7 emerged from a resilience assessment of the Goulburn-Broken catchment in southeastern Australia (Walker et al. 2009). The Goulburn-Broken catchment is a productive agricultural region facing a range of resilience-related issues. It featured in the book *Resilience Thinking*.

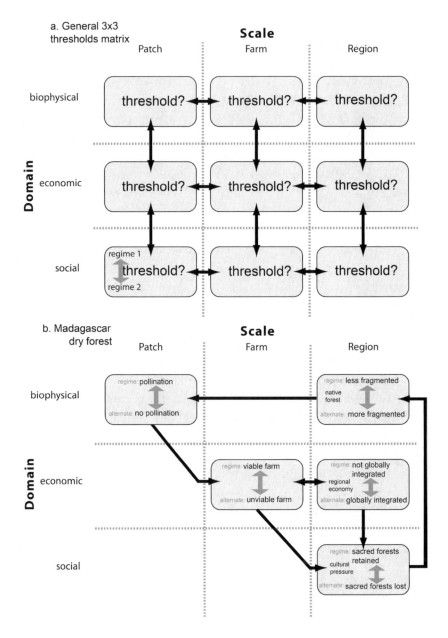

Figure 6: A Representation of a Thresholds Matrix

(a) A generalized 3x3 grid of potential thresholds over three scales and three domains. Each box identifies a potential threshold on a controlling variable that leads to a regime shift. (Modified from Kinzig et al. 2006). (b) How researchers believe thresholds might be interacting in a system of dry forests in Madagascar. (Based on Bodin et al. 2006, modified from Kinzig et al. 2006.)

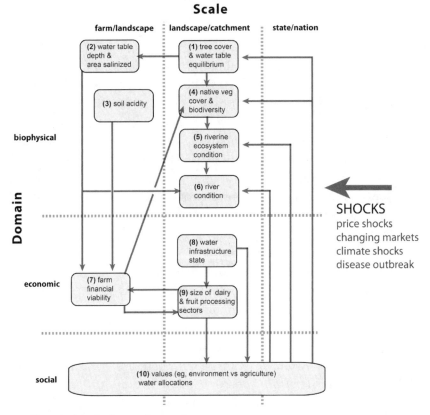

Scale

farm/landscape   **landscape/catchment**   **state/nation**

Domain

biophysical

(2) water table depth & area salinized

(1) tree cover & water table equilibrium

(3) soil acidity

(4) native veg cover & biodiversity

(5) riverine ecosystem condition

(6) river condition

SHOCKS
price shocks
changing markets
climate shocks
disease outbreak

economic

(7) farm financial viability

(8) water infrastructure state

(9) size of dairy & fruit processing sectors

social

(10) values (eg, environment vs agriculture) water allocations

Thresholds & alternative regimes

1. **Landscape tree cover** (10 - approx 15%) water table deep vs. water at surface
2. **Water table depth** (approx 2m under surface) fertile top soil vs. salinized soil
3. **Soil acidity** (approx pH 5.5) high crop production vs. crop failure
4. **Native vegetation cover** (5%, 30%) fauna groups present vs. absent
5. **N and P content in water bodies** (concentration level threshold) clear water, high biodiversity vs. eutrophic, low biodiversity
6. **River flow regime** (flow level threshold) wetland/floodplain biodiversity persists vs. different species
7. **Farm income:debt ratio** (debt ratio threshold) farm viable vs. non-viable
8. **Funds for infrastructure from water sales** (level of funds threshold) infrastructure good vs. declining, non-viable
9. **Size of dairy & fruit processing** (level of industry activity) viable processing plants vs. non-viable
10. **Balance of values for water for environment vs agriculture** (availability of water threshold) farmer income: debt positive vs. negative

Figure 7: A Thresholds Matrix for the Goulburn-Broken Catchment

The matrix presents ten slow variables with identified thresholds. The arrows between boxes indicate possible cascading threshold effects. The list provides more details on each threshold: each item names the slow variable, the threshold in parentheses, and the regime and alternate regime that lie on either side of the threshold. (Modified from Walker et al. 2009.)

In the Goulburn-Broken catchment, assessing specified resilience involved identifying the controlling variables that might have threshold effects. Consider threshold 3 (soil acidity) as an example. As soil acidity increases (pH declines) from normal levels, crop production is not affected very much. However, when pH drops below about 5.5 there is a sudden drop in the ability of crop plants to grow, because below this pH level, aluminum in the soil becomes soluble, enabling it to be taken up by the plants. Aluminum is toxic to plants and inhibits their development. The controlling variable is the pH (acidity) of the soil, and the threshold effect occurs at pH 5.5.

This threshold in soil acidity is a good example showing how thresholds occur as a result of a change in a feedback process. The regime shift is from high crop production, at high pH with no aluminum toxicity, to low crop production, at low pH with aluminum toxicity. Soil pH declines when nitrogen is added to the soil. It's an unintended consequence of fertilization, and the critical step in the feedback process is the effect of pH on the solubility of aluminum. At a pH of 5.5, aluminum becomes very soluble. At this point, there is a critical change in the effects of this feedback process from nitrogen fertilization, to soil pH, to aluminum toxicity, to crop production.

Threshold 2 relates to the level of groundwater. As the water table in a landscape rises (a process that is connected to tree clearing at the scale above), the productivity of the soil remains pretty much the same until the water table reaches a level of about two meters under the ground surface. This varies a bit depending on soil texture but it's roughly two meters. When it reaches that level, the water gets drawn up to the surface by capillary action driven by evaporation of water from the soil surface. The water, and any salt dissolved in it, then fills the topsoil profile. Following this, plant growth (crop production) plummets.

As discussed in chapter 1, this is effectively an irreversible threshold in soils with any significant amount of clay. Even if the water table drops down again below two meters, the system takes a long time to go back to what it was, because the salt (sodium) causes dispersal of clays, making the soil soapy and "claggy." There is much-reduced percolation. It takes a long time and lots of rain to flush the salt down to below the root zone. The controlling variable is the water table depth, and the threshold occurs at around two meters.

The participants in the Goulburn-Broken assessment were concerned about a complex threshold effect operating in the social domain over all three scales. It had to do with whether society places a greater value on agricultural production or environmental functionality. At the time of writing, there is a fierce debate raging over this very issue; and this debate is occurring at multiple scales. Which should have priority in the allocation of water: environmental flows or production? The production people fear a tipping point in public opinion that will shift the allocation in favor of the environment, thereby increasing the cost of water for them, and the amount available. Some believe it has already tipped, as evidenced by national government policy development, but it is being strongly resisted by state governments and regional government organizations, and the outcome is not yet clear. It's a debate that is also playing out around the world as available water resources are facing increased demand and a growing list of threats.

As with the Madagascar example (figure 6b), the thresholds in the Goulburn-Broken catchment are of different types. Some are stepwise thresholds, others have hysteretic pathways, and some are irreversible. This produces different forms of interactions and cascading impacts.

Figure 7 came from a region in southeastern Australia in which agriculture is the predominant way of life, though other goods and services are also highly valued. Equivalent diagrams for a coastal region in which fishing is the dominant way of life, or a boreal forest region in which timber harvesting is predominant, would look somewhat different. In some assessments participants have preferred to combine the social and economic domains and use a socioeconomic and a biophysical domain, in a 2 domains × 3 scales matrix. The key thing is to examine the different domains at a focal scale and a number of other scales at which the system functions. And in each combination it is likely there would be actual or potential threshold effects on controlling variables.

It can also be helpful to conduct separate assessments in a region, using different focal scales, one embedded within the other. An example (in case study 3) is the assessment of the Macquarie Marshes as a focal scale, as well as the larger catchment region within which it sits. Both used their own focal scales (together with finer and higher scales) and respective sets of thresholds and cross-scale influences. And each assessment informed the other.

Tipping points is another name for thresholds, although the term is used mainly in association with social systems. Tipping points have been recognized for a long time, and a body of theory has developed around them—see, for example, Granovetter (1978), who describes threshold models for the behavior of crowds involved in rioting. Controlling variables in social systems can change much more quickly than those in ecological systems. In fact, sometimes they change suddenly (e.g., social preferences, voting intentions, market demand). So, although many of them do change slowly (population age structure, religious tolerance), in social systems the relative speeds of variables often make the distinction between "slow" (controlling) and "fast" (response) variables not all that useful. Also, new things crop up in social systems, technologies, and social phenomena (like YouTube and Twitter) that have no parallel in biophysical systems.

Because of all this, social system thresholds in social-ecological systems are difficult to determine, and some believe it's not profitable to pursue the idea. However, if they do occur it is important to know about them. And there is enough evidence for their existence to warrant giving them careful consideration. One interpretation of recent upheavals in the Middle East suggests that there was a tipping point in some complex variable of social inequity and level of unemployment. Knowing about the existence of thresholds of this nature makes a big difference in how policy and strategies for social-ecological systems (nations) are developed.

We come now to the big question: How do you identify thresholds in your system? How do you come up with a thresholds matrix along the lines of figure 7? In the following discussion we suggest four ways this task might be approached. Each successive method is more demanding in what it requires. To begin with, it's helpful to consider known thresholds and thresholds of potential concern. Then it's a matter of attempting to construct simple conceptual models of how your system is operating. Finally, you might consider engaging experts and developing analytical models.

1. Known Thresholds

Are there any known thresholds in the system? They may have already been crossed in some places within the system, or they may be known from other systems that are very similar.

If two meters is a threshold for the depth of saline groundwater in the Goulburn-Broken catchment, then it's likely that a similar threshold, depending on soil type, will apply to other irrigated catchments in the region faced with the threat of rising groundwater.

Similarly, soil acidity, nutrient levels, fire frequency, and grass cover have all exhibited threshold effects in some systems, so they are good starting points for discussion in your system if these variables affect the flow of ecosystem goods and services.

If your system is anywhere in the semiarid tropics, and especially if it has sandy soils, then it will almost certainly have a threshold for soil organic matter content, below which soil fertility declines very significantly. And this is often related to length of time an area is left fallow. In the western African country of Niger, as described by Hiernaux and colleagues (see Fernandez et al. 2002, pages 313–318), a threshold value of less than three out of eight years fallow in semiarid areas leads to declining soil organic matter, and hence desertification (as opposed to the maintenance of soil organic matter with higher ratios of fallow years to cultivated years). It will probably not be exactly 3:8 in other regions, but it is very likely that a fallow:cultivated threshold ratio close to this will exist.

In an analogous way, if your system is a Northern Hemisphere lake, then it is almost certain there will be a threshold amount of phosphorus in the lake sediment, above which it is inevitable that the lake will flip from a clear state to a murky, algae-dominated state (Carpenter 2003).

There is a developing typology of thresholds on the website of the Resilience Alliance (www.resalliance.org) that is a helpful point of departure, and it is also useful to consult the regime-shifts database being developed in the Stockholm Resilience Centre (http://www.regime-shifts.org/). It describes in some detail examples of regime shifts, with descriptions of the threshold effects involved.

Thresholds in the social and economic domains might be harder to identify. We've described how tipping points in the social domain are well known in terms of their behavior, but they are context dependent and it's unlikely that you will be able to infer the existence of particular tipping points of this kind in your system from known examples elsewhere. In the economic domain there are more repeatable examples, and a good place to begin looking for them is in systems similar to your own, facing similar trends. For example, when do levels of debt become unserviceable at the farm scale? In your region, is this a serious issue that needs to be included? When will declines in population in regional centers make them undesirable places for new businesses to establish themselves, and is there a threshold level for this? The degree of uncertainty involved in social-system dynamics suggests that suspected thresholds fall more into the next category.

2. Thresholds of Potential Concern

Some thresholds may be suspected rather than known. But this doesn't stop you from incorporating them into management.

The notion of "thresholds of potential concern" (or TPCs) was first developed by the managers of Kruger National Park in South Africa, where it was introduced in the mid-1990s. They did not have the resources to monitor or manage everything in the park, so they set about trying to identify the really important things that they needed to know about and manage. They came up with the idea of TPCs, in which they focus their research, monitoring, and management on an evolving list of actual and potential thresholds that together determine the trajectory of the park (Biggs and Rogers 2003).

In workshops with managers of agricultural regions in Australia, we have found TPCs to be a helpful tool in getting to a first-cut assessment of possible thresholds, a means of identifying candidate threshold effects. The Macquarie Marshes assessment in case study 3 is one such example.

The Kruger National Park researchers and managers use TPCs as monitoring end points for managers. When a TPC is reached, management intervention is needed. In this sense, TPCs are a set of operational goals. They were originally defined as the upper and the lower levels along a continuum of change in selected environmental indicators for which the Kruger ecosystem is managed. For woody vegetation cover, for example, the TPC would be exceeded when woody cover for any landscape dropped by more than 80 percent of the highest-ever value (Gillson and Duffin 2006). As a part of an adaptive management approach, TPCs are being continually adjusted in response to the emergence of new ecological information or changing management goals.

TPCs in Kruger are now being revised from a concept of upper and lower limits to one which operates in terms of rates of approach to thresholds. Rates of approach provide an indication of how fast the system is moving toward a point of undesirable system change. This allows managers to build in confidence buffers and to plan management action, acknowledging it takes time to respond.

The experience gained from trying to operationalize TPCs has led the Kruger people to a revised definition of a TPC, one that incorporates the social part of the system, explicitly recognizing the subjective, value-determined equivalent of a biophysical threshold. They link it directly to management/policy decisions.

The idea of using both biophysical and human-determined "utility" thresholds is also an integral part of what Martin and colleagues (2009) call "structured decision making." As they explain it, ecological thresholds are determined by our understanding of the ecology of the system and are incorporated into models of system behavior. Utility thresholds, on the other hand, are determined subjectively and reflect stakeholder values. They point out, however, that in some circumstances these values can be based on knowledge of the ecology of the system. The utility thresholds result in setting objectives, based on human values. The two sets of thresholds (ecological and utility) are brought together to arrive at "decision thresholds."

3. Developing Conceptual Models

A "mental model" of operation is how someone believes a system works and changes. All stakeholders have their own mental models on how things work in their system, though many people may not realize or even acknowledge it. Getting these mental models on the table, and shared, is a necessary start to this part of the process.

It's inevitable that there will be different mental models among any group of stakeholders. Rather than trying to detail them all, start by attempting to develop an agreed-on, explicit conceptual model, noting where differences might arise. See if any of these differences can be resolved. Parts of the system operate in a manner that everyone knows and accepts. People might have different ideas on some of the relationships, and it's important to note these, but try not to get bogged down by the differences.

What about critical levels in important variables? Will the enterprise, ecosystem, community begin operating in a different way if certain levels are crossed?

At the simplest level, figure 8 illustrates the kind of mental model exploration that can help get you toward identifying thresholds. It in-

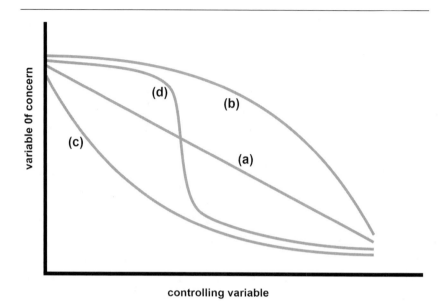

controlling variable

Figure 8: Patterns of Response

Possible patterns of response in the state of a system (variable of concern) to changes
in the amount of the variable that determines its state (controlling variable): (a) linear,
(b) and (c) still smoothly changing but curvilinear, (d) step change or threshold effect.

volves questioning the shape of the response of the variable you are
concerned about (some valued state of the system, or system service)
to its underlying controlling variable. If it's linear, as in (a), there is no
critical level. If it's more like (b) than (c), it is "safer" in the sense of
not declining as quickly as the controlling variable increases. Or could
it be like (d), which would suggest a strong step change or a threshold
effect? It is surprising how a discussion of this kind reveals different
mental models of how the system works.

One useful way to move ahead is to try to develop an agreed-on
"state-and-transition" (S&T) model for the system. S&T models were
originally proposed by ecologists for rangeland systems in the late
1980s (Westoby et al. 1989). This approach proposed that rangeland
dynamics can usefully be described by a set of discrete "states" of the
vegetation on one piece of ground and a set of discrete "transitions"
between states. For each transition there is a set of conditions under
which it can/will occur.

S&T models force people to be explicit about how they see the system functioning, and they highlight differences that then need to be resolved. There are now many examples of S&T models, ranging from very simple box-and-arrow diagrams to quite sophisticated quantitative models. Rangeland professionals in the United States have developed a suite of S&T models and use them routinely (see Briske et al. 2008). If you are interested in pursuing this further, look at Rumpff et al. 2011, Bestlemeyer et al. 2009, and Suding and Hobbs 2009.

At various scales and in different domains (biophysical, social, and economic), or even including more than one of each, can you describe the current state of the system and the possible alternate states it could be in? As an example, figure 9 came out of an attempt to describe the dynamics of the Camargue region in Southern France.

The Camargue is the delta of the Rhône River valley. For centuries arguments have raged about how different parts of it should be managed. Those practicing the ancient traditions of reed harvesting for thatching have attempted to get as much of the region as possible into the state that favors reeds. But this is also the region of the famous Camargue cattle and white horses, and they need meadows and grassland. The region is also much in demand by duck hunters who would like lots of open water, making it good habitat for ducks.

Many dikes or banks have been constructed over the years to control the water, because water is the major factor determining the state of any area in the Camargue. The reeds, being a flood-tolerant grass, prefer inundation, but a summer drawdown every five or ten years improves the stability of the reedbeds. Permanent flooding results in the progressive spread of open water at the expense of reeds. In contrast, very frequent and long-lasting dry periods result in the encroachment of woodland.

This example shows how different ecosystem services are linked, as discussed in chapter 2. The reedbeds provide fish hatcheries and reeds for thatchers, the open water is good for ducks, and the wet meadows are good pastures for livestock and horses. Increase the level of one service, and you reduce the availability of others.

The possible alternate states of the region are shown in figure 9, together with the conditions that are needed to bring about transitions from one state to another. Some of these may have threshold effects on them. For instance, those from "forest" to

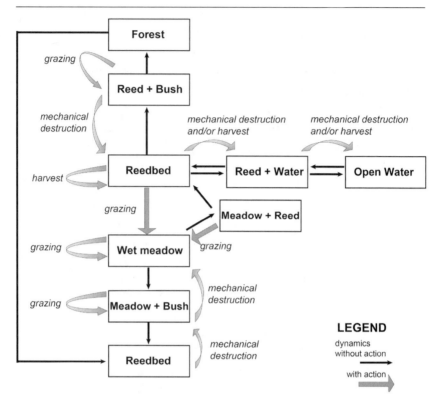

Figure 9: A State-and-Transition Model
of the Camargue Wetland System

Boxes represent alternate states the system can be in. Arrows define how interventions
bring about transitions. (Modified from Mathevet et al. 2007.)

"meadow + bush" and from "meadow + bush" to "wet meadow" require mechanical destruction to bring them about. The transitions
in the other direction come about through internally driven processes, but they are not reversible. They therefore represent irreversible thresholds, from a systems perspective. Reversing them
requires mechanical intervention.

Where and how you start developing your own S&T model depends
on the information available. To begin you might try asking, What are
the possible states the system can be in? Some of these will be known
because different parts of the region may be in different states, or because the same kind of system is in, or has been in, different states in

other areas or at other times in its history. Some of the possible states may be guesses based on what people think could happen.

It is useful in the beginning to consider all sorts of possible states, but the aim is to try to arrive at a limited set of states that all agree are possible and have significant consequences.

Next, what transitions between these identified states are possible, and what are the necessary conditions for the transitions to occur? Can you identify possible trajectories for the system? For these trajectories, can you identify the different end states the system could be headed for and what the intermediate states might be? In doing this, you need to think about the states of the system in terms of both the ecological and social domains.

Along the various pathways, are there any no-return points, or thresholds? The important point is to consider the possible transitions between the different states (noting that not all states can lead to all other states) and then to ask, Could any of these transitions have threshold effects? Are any of them nonreversible?

A note of clarification: these S&T models superficially resemble the threshold matrices (as presented in figures 6 and 7); they are both diagrams consisting of a set of boxes linked by interacting arrows. But they are quite different. The boxes in the S&T models identify different states the system can be in at any one place. The arrows indicate how that site might change (transition) from one state into other states. In the thresholds matrix, each box identifies a threshold shift from one state of the system to another. It represents two alternate states and the threshold separating them. The boxes are configured in an array based on the domains and scales that make up a system. The arrows between boxes represent interactions between thresholds. S&T models are a good way to envisage or highlight possible thresholds. They are one means of getting *to* a thresholds matrix, but the two frameworks are quite different.

Developing a version of a thresholds matrix using TPCs and conceptual models, and identifying the interactions among the thresholds, is in essence giving you a picture of specified resilience in your system. Keep in mind that developing these models is always a work in progress. The model always needs updating and is largely an instructional aid for discussing and understanding the specified resilience of your system.

4. Analytical Models

Beyond these three methods of developing a thresholds matrix (i.e., known thresholds, TPCs, and S&T models), you will probably require involvement of specialists and scientists to help develop explicit analytical models of the alternate regimes and their defining thresholds. The following section is accordingly somewhat technical (with a few more references). It is offered to demonstrate the types of approaches that you might consider using but that require specialist input. Use it as a guide but don't be put off if the finer points are a bit obscure.

As a kind of segue into these more advanced quantitative models, you might consider trying what is known as fuzzy cognitive mapping. This technique can assist in the development of scenarios for environmental management. Kok (2009) and Özesmi and Özesmi (2003) provide good accounts of how it is done. It's a semiquantitative tool that makes you think about the important components and issues, or drivers, in the system.

In fuzzy cognitive mapping there is a central issue of importance—like deforestation in the Amazon (Kok example) or lake pollution (Özesmi example)—and the aim is to get a common understanding about the important things (concepts) that influence this central issue. The model consists of (1) a vector of these concepts, the values in the vector reflecting the importance of each concept in regard to the overall issue being considered, and (2) a matrix of their effects on each other, basically from -1 to $+1$ in each cell of the matrix (with several likely to be "0"). The vector is multiplied by the matrix to give a new "state of the system" (new vector values), and this is repeated until the values either stabilize or implode, or explode. The presence and value of feedbacks are all important in the outcome, and hence its value in helping to develop a coherent "model" (understanding) of how the system works.

Fuzzy cognitive mapping sits in between identifying issues in the system and modeling how they work.

Dynamic Models

There are all kinds of models and modeling approaches that can shed light on the issues you identify. The most common are quite simple mathematical models of linked differential equations.

One way to develop them is by making S&T-type models very explicit by writing the equations of motion for each of the important (defining) components of the system. In systems terminology, these are called state variables because the amounts of each of these describe the state of the system. Examples are the models by Anderies and colleagues (2002), for exploring the positions of grazing thresholds in a rangeland, and by Carpenter and colleagues (1999) for exploring regime shifts in lakes due to eutrophication, and how the positions of the thresholds depend on the composition of the fish. The value of these models is in exploring the sensitivity of the system to the various parameters that determine how the variables in the system change.

The process of building the model forces you to be very explicit about how things change and the effects that a change in one component will have on other components. It tests your understanding of the system and does not allow you to be vague and imprecise. In some cases you may have an idea about a possible threshold, so you can construct the equations to test whether it can occur. The model will tell you what the values of the variables and parameters have to be for the threshold effect to occur. In other cases the failure of the model to produce credible results will force you to consider what is missing and to consider nonlinear (threshold) effects that may be responsible.

Thresholds are always associated with a change in a critical feedback process, and to explore feedbacks you need an understanding of the mechanisms underlying the threshold effect. Table 2 shows the feedback changes involved in published accounts of thresholds on a range of controlling variables, based on analysis of the Thresholds Database described in Walker and Meyers (2004). Each process listed under "associated feedback changes" comes from a published example in which a critical change in that process, occurring at a particular level of the controlling variable, resulted in a change in the trajectory of the system (and hence a regime shift).

The position of the threshold on the controlling variable occurs where a significant change in the feedback process happens. Managing for resilience therefore amounts to managing for feedbacks and knowing where the feedbacks change.

Understanding feedbacks requires a mechanistic understanding of the system's dynamics. The following two examples of counterintuitive feedback effects provide a good insight into how they can work.

In lakes in Papua New Guinea an introduced aquatic fern, *Salvinia*

Table 2. Feedback Changes and Thresholds

Controlling variable	Associated feedback changes
Rainfall	Evapotranspiration, leaching, water table level
Temperature	Soil moisture (evapotranspiration), germination (micro-climate), symbiosis (coral bleaching)
Nutrients	Oxygen in water (decomposition), competition (plant species)
Acidification	Calcification (plankton)
Vegetation amount	Water interception (cloud forests), infiltration rates, water tables, nutrients (legumes), soil temperature (insulation)
Percent native habitat in landscape	Immigration/emigration rates, reproduction, survival
Herbivory	Regeneration, competition, fire (fuel)
Harvesting	Recruitment (depensation), evapotranspiration (forests)
Frequency of fires and fallows	Seed bank viability, regeneration time
Predation	Recovery (depensation), herbivore behavior (due to spiders, wolves)

Though not included in the published accounts, we add two more examples for which there seems to be evidence of feedback changes leading to regime shifts.

The economy	Income: cost ratios, debt:income ratios
Social preference (e.g., water for environment or for agriculture)	Subsidies/taxes

molesta, has caused enormous problems for native species and lake users by forming massive floating weed mats. Similar problems in other tropical countries have been solved by introducing a small species of weevil (*Cyrtobagous salviniae*) that feeds on the fern. Unfortunately, in Papua New Guinea the introduced weevils just died out. The solution—though it seemed like a crazy idea—was to fertilize the salvinia to make it grow more.

Ecologist Peter Room found that nitrogen levels in the salvinia were too low to provide sufficient protein for the weevils to breed (Room and Thomas 1985). When he fertilized salvinia in caged treatments, the weevil population took off and did the job. When he released the weevils on wild salvinia mats, he found that if the density of weevils was high enough, they managed to breed even without fertilizer because insect damage to the salvinia increases its nitrogen content.

A critical level of damage (and therefore density of insects) was needed for this to happen. Below that threshold density, the insects died out. Above it, they multiplied. The critical feedbacks in the system were from salvinia food quality to the weevil breeding requirement and from the level of weevil damage to food quality of the weed. Bringing about a change in the feedback processes led to a regime shift in the *Salvinia-Cyrtobagous* system.

In another example, counterintuitive responses of fish populations to management have been recorded in a number of cases (Pine et al. 2009), and each has been due to unexpected behavioral responses and changes in juvenile survival rates of the fish. A brook trout example from California illustrates this. Introduced brook trout (*Salvelinus fontinalis*) successfully spawn in alpine lakes in California and overpopulate, causing a range of problems including reduced size of adult fish. Based on conventional wisdom, Pine and colleagues reasoned that lowering trout densities would mean more food for adult trout and so would lead to improved angling.

However, when they experimentally removed adult trout, reducing the density by 50–80 percent, there was either no improvement or reductions in trout growth. There was, however, a dramatic improvement in the survival of young trout (up to 1 year old). This was apparently due to reduced cannibalism by larger trout. The resulting large juvenile cohorts spread all over the lake, rather than being confined to edge areas (where they previously sought refuge from adults), competed with adults for food, and negated the expected improvement in adult fish growth. The management "model" did not take into account the feedback from adult trout density to juvenile survival and distribution.

Changing Patterns in Time and Space

It's often stated that it's impossible to detect a threshold until it's been crossed. But that's not true. Many systems, as they approach a threshold, start changing their dynamics. Their variability increases. They start to fluctuate more than usual, and this "rising variance," as it has been termed (Carpenter and Brock 2006), has been used as a leading indicator of a pending regime shift. In their example of lake eutrophication, Carpenter and Brock showed how the variable that causes the regime shift—the amount of phosphorus in the water—started to increase in variability before a regime shift, signaling an impending shift about a decade in advance.

In addition to changes in variance, another time-related change has to do with the tendency of the system to return to its stable (equilibrium) state. As a threshold is approached, this return time to equilibrium increases (Wissel 1984). It's called "critical slowing down" and shouldn't be confused with changes in variance. Increasing variance is about fluctuations in response to environmental variation (random "noise" in the environment); slowing down is about the time the system would take to return to its stable state at a given amount of the controlling (slow) variable (have a look at figure 1 and consider the arrows showing how for any level of the controlling variable, the system moves toward a stable state). And as the amount of that slow variable gets closer to the threshold, the time the system would take to return to equilibrium increases.

The important thing that follows from these two properties (rising variance and critical slowing down) is that as a system approaches a threshold, autocorrelation increases (today starts to look more and more like yesterday) and so does the variance. These concepts are not all that easy to grasp, and an excellent book on thresholds and how to model them is *Critical Transitions* by Marten Scheffer (2009). It and a pair of papers by him and his colleagues (Dakos et al. 2008 and Scheffer et al. 2009) explain how autocorrelation and rising variance work and are expressed in different kinds of systems. They have used models tested on data from a number of different systems to explore how you can know when a system is approaching a threshold.

In a way analogous to changes in time, changes in spatial variation have been shown to indicate approaching thresholds. It has been nicely demonstrated in arid grazing systems by Rietkerk and colleagues (2004), who describe how self-organized patchiness that characterizes "healthy" and productive arid rangelands (remember the tiger stripes pattern discussed in case study 1) loses its intensity of pattern and collapses to a homogeneous, low-production state as average grazing intensity (the slow variable) increases. Examples of increasing autocorrelation in space (patch sizes get bigger) as thresholds are approached are described in Dakos et al. (2010) and in Scheffer's book.

So, three features of the behavior of the "fast" variable (that displays a regime shift when a threshold is passed) seem to repeat in different systems as a threshold is approached: (1) variance increases, (2) autocorrelation increases (critical slowing down—today looks more and more like yesterday), and (3) spatial autocorrelation increases—bigger patches of the variable develop.

These ideas, and their associated statistical methods, are still being developed, and until now their use has been limited by their requirement for data. This is especially the case in regard to the two time-based measures. For many systems, the time required to statistically detect an increase in variance means the system will go over the threshold before detection is possible.

The spatial autocorrelation results, however, are more promising because modern technology can collect large amounts of spatial data for many systems at one time. This allows the detection of the pattern type/intensity that could indicate the point of a regime shift.

The science of early warning signals is an area of rapid growth. At the time of writing, two examples are worth noting. Steve Carpenter and Buzz Brock are soon to publish work on the use of statistical fitting techniques that use the three components of change in a variable as it approaches a threshold. Specifically, they use a "drift-diffusion-jump" model, where drift measures the local rate of change, diffusion measures relatively small shocks that occur at each time step, and jumps are large intermittent shocks. Quite separately, work in progress by J. Carstensen and A. Weydmann on Arctic change is demonstrating that changing variance could have been used some time ago to signal the loss of Arctic sea ice. Using remote sensing data, they suggest that the loss rate of sea ice accelerated by a factor of about 5 in 1996, but increases in random fluctuations, as an early warning signal, could be observed in 1990.

As we have said, using analytical methods such as these for the detection of thresholds comes later in the assessment process, is very context dependent, and goes beyond our aims in this book. The above introduction, however, should help those interested to get started. The following notes introduce a couple of different modeling approaches.

Network models: These are models of how all the variables in the system are connected to each other. One well-known property of networks is that as links are added at random between the members of a network, there is a gradual increase in connectedness, with "loops" starting to appear at moderate levels of linkages. After that, however, the addition of the next few links can very suddenly result in a huge increase in connectedness, with almost all members being included. Network models can indicate critical weak points, or change points, especially for the social parts of the system.

All ecosystems and social systems have network structures, and examples of sudden changes because of network properties have been recorded in each. They are particularly important in social systems, because they are more likely to change in social systems. If the following kinds of questions apply to your system, it may be worth developing a network model (a more complete list and ideas on how to develop a network model are given in the Resilience Alliance's *Workbook for Practitioners*, available at www.resalliance.org/3871.php).

- Are there key people or groups who are not connected to others but who are affecting the potential for solving resource issues?
- To what extent do highly central (connected) and potentially influential actors represent the views and interests of the other stakeholders? If centrality is a strong feature of the network, is it a source of social cohesion or a potential barrier to achieving it?
- Are there any actors in the network that link otherwise separated groups, and do they represent bridges or barriers to collaborative governance?

Agent-based models: These are like gaming models that allow you to explore the outcomes of various policy decisions under a range of assumptions about how people react to changes in both ecosystems and governance. They involve defining a number of "agents" (like all the farmers in a region) and assigning to each "rules" for managing their resource.

For example, how much rain should fall before you decide to plant a crop? Each agent is randomly assigned a different amount of rain for making this decision, the amount being randomly chosen from a range of rainfall amounts. The model has embedded in it a model for crop production that has all the essential processes that are involved. The agent-based model is run under various rainfall "seasons" repeatedly, and the agents who emerge as most successful define the sets of management decisions (rules) that are best for that region.

Developing such a model for rangeland management, using the basic mathematical model of rangelands dynamics referred to earlier (Anderies et al. 2002), allowed for interesting insights into the need for learning through mistakes and into how to avoid undesirable outcomes (Janssen et al. 2000).

A well-developed approach involving agent-based models and stakeholder participation is being used by a group based in the French agency CIRAD in Montpellier, France (Bousquet et al. 1999). It is called ComMod (short for companion modeling) and provides a modeling platform for engaging with stakeholders and then enabling them to develop their own model of how their system works, including its management. By including rules for using resources, based on different agents' views and judgments, the stakeholders are able to see how their system changes in response to their decisions. Bousquet and his colleagues have successfully used the approach in a number of developing-world regions. See www.commod.org/en for more information.

Assessing Specified Resilience

Specified resilience is assessed by identifying alternate states and associated thresholds. In the sections above, we've outlined four categories of engagement with specified resilience. Some might consider them steps:

1. Known thresholds: List what's known.
2. Thresholds of potential concern: List what's suspected.
3. Conceptual models: Make explicit your (shared) understanding of how the system functions, and use it to identify possible thresholds.
4. Analytical models: Flesh out the conceptual understanding with quantitative measurement.

While the best appreciation of your system's specified resilience might involve all four steps, progress in any of them is improving your engagement with resilience. And the best outcome is achieved by going through these steps iteratively as new information and understanding emerge.

As a final comment on specified resilience and thresholds, we urge you to bear in mind that what you are trying to achieve is some version of the scales-and-domains framework of thresholds, and it helps to recognize that there is a hierarchy of threshold effects, with some thresholds embedded within the effects of higher-scale, or more dominant, thresholds.

In the Australian examples (case study 3, after this chapter) we describe some of these, and the Macquarie Marsh assessment highlights this hierarchy effect—some thresholds are dependent on what happens to others, but not the other way around. For instance, a threshold for the minimum dispersal distance for native fish species (needed for their persistence) only exists as long as there *is* a marsh. If the flood regime threshold for the core marsh is passed, then native fish species will be lost anyway. But if the fish dispersal threshold is passed and some species disappear, it does not directly influence the existence of the core marsh. Though the fish threshold is important in its own right, the hierarchical nature of the thresholds makes management of the marsh flood regime a higher priority for policy development. Getting such a hierarchical picture of the thresholds is therefore helpful, as it simplifies understanding what to do about it all. It leads to a hierarchy of management interventions.

Clearly, from everything we've just discussed, assessing specified resilience is far from a trivial challenge. Unfortunately, it's also not the end of the story. Because you can never be sure you have identified all thresholds, and because making your system resilient in particular ways can cause it to become less resilient in other ways, you need to also consider and incorporate general resilience in your assessment. In many ways this is a bigger challenge because, while resilience scientists have put much effort into understanding and working with specified resilience, the field of general resilience is only beginning to open up.

General Resilience

While assessing specified resilience in a system, it is important not to forget what might be happening that we are not directly focusing on. For example, is it only the resilience of crop production to drought that is of concern? What about the resilience of other ecosystem services to other shocks? And what about the resilience of various parts of the social system? These types of questions lead us to a consideration of general resilience—the capacity of a system that allows it to absorb disturbances of all kinds, including novel, unforeseen ones, so that all parts of the system keep functioning as they were.

Another way of putting this is by asking how your system—a business, family, farm, national park, or whatever system you have some responsibility for—would cope if things got really tough. We're not defining any particular aspect of the system, and we're not saying what the specific problem/disturbance is—we're talking generally.

If you're the mayor of a village in a fire-prone area and you're asked about preparedness for fires, you'll consider things like firefighting capacity, knowledge of fuel loads, fire risk assessment, and a raft of things specifically to do with being prepared for fires. But if you're asked how prepared the village is to cope when an unknown disaster strikes—it could be a disease outbreak, flood, earthquake, riot, gas explosion, or something right out of left field—then your answer will be more about the general qualities of the village. What are its food reserves, diversity of skills to deal with different types of emergencies, levels of trust and the ability of the community to pitch in to help itself, distance from the nearest hospital, friends in high places who will mobilize resources to help when things get tough, and so on. These are all general qualities and they all relate to general resilience.

General resilience is related to adaptability (or adaptive capacity), since the attributes conferring both largely overlap. And those attributes include features such as diversity, modularity, the tightness of feedbacks, openness, reserves, and high levels of all the types of capital (financial, human, natural, built, and social).

In a broad sense, a resilient system is forgiving of mistakes (in policy and management) and it can absorb large shocks. General resilience therefore has three important functions:

- Being able to *respond* quickly and effectively, in the right places in the right way
- Having *reserves* and access to needed resources, thereby effectively increasing the "safe" space for operating
- Keeping *options* open

A resilient pastoral enterprise, for example, would have a greater capacity to recover from the management mistake of not destocking sufficiently going into a drought. A resilient fishery might cope better if fishing quotas were set too high. In a less resilient system, both mistakes might shift the system into an alternate regime that might be difficult, or impossible, to return from.

Assessing General Resilience

Various studies around the world have identified diversity, modularity, the tightness of feedbacks, openness, and reserves as important for general resilience. They have shown that in various circumstances one or more of these attributes have been critical to sustaining resilience, but this doesn't mean that they are all important in all situations. The list, then, is merely a starting point to guide thinking. Which of these and which other attributes are important will depend on the system being considered.

The theory behind this area of general resilience is not well developed and needs, among other things, a retrospective comparative analysis of sets of case studies. These case studies need to operate over different spatial scales, domains, and time scales. Such a comparison would help us elucidate the attributes of general resilience, where and when they apply, and the interactions among them.

There is a lot of overlap between general resilience and the notion of robustness. *Robustness*, however, has more of a design connotation. For example, how do you design a bridge, or a management policy, that will continue to function under a range of conditions? You want a system that is robust to changes in the environment—biophysical and socioeconomic. From a resilience perspective, however, this is a problem because the analysis of robustness requires specifying the range of uncertainty (conditions) the system must be able to cope with. It doesn't allow for novel shocks—the famous "unknown unknowns." The work on robustness is, however, a valuable contribution to understanding and dealing with general resilience.

In terms of their influence on the resilience of the system, all the attributes of general resilience interact. It is therefore not possible to determine the amount of any attribute that marks a critical level. It depends on the amounts of all the other attributes. This makes it difficult, if not impossible, to quantify general resilience in absolute terms, and the most appropriate approach is to try to identify trends and changes and examine them in terms of possible effects.

As an example, consider the attribute of "diversity." A full list of all the kinds of diversity in any system—social, ecosystem, and infrastructure—becomes very long (skills, age structure, ethnicity, types of farming enterprises, crop types and varieties, employment, transport options, habitat, species, and so on). Such an effort is more likely to hinder understanding. So, rather than developing a long list and trying to determine the levels of each attribute, it is better to start by considering whether there have been any changes in the system that could influence its capacity to cope with a shock. It's best to do this in an iterative way, by referring back to the description of the system— in particular, the big issues. Based on your developing model (understanding) of the system, could any of the changes that have occurred or are occurring have significant effects on resilience, in general? Which of these warrant collecting more information? Develop a working list of system components/areas where trends may be of concern.

The following notes on each of the attributes will help get you started.

Diversity

Have there been any changes in diversity that might relate to the valued goods and services identified earlier? A simple example is the change from multicropping to monocropping in agriculture, or a decline of bird species in the area, or the general aging of the community with many older people but no young families coming into the area. Do any of the changes amount to persistent trends, and could any of these make the system more vulnerable to external shocks? In which parts of the system is there little or no diversity, and does this make the system vulnerable?

In terms of resilience there are two forms of diversity worth noting: functional diversity and response diversity. *Functional* diversity refers to the different functional groups of organisms that are represented in an ecosystem, or different functional groups of people in the social domain. Different functional groups do different things. In an ecosystem, one group might fix nitrogen while another might assist in the breakdown of waste. In the social domain, you might have doctors, lawyers, and engineers, each providing a different class of service.

Within each functional group there is usually a range of species, or doctors or engineers and so forth, that provide the same basic service, though they go about their business in slightly different ways. For example, in coral ecosystems there might be a number of different fish species that graze on fleshy algae, and they each have their own abilities to tolerate different kinds of shocks and disturbances. In the social domain you might have a range of different types of engineering companies; some might be small specialist groupings undertaking specific work, and others might be large companies capable of taking on a broad spectrum of tasks.

The important point is that the different species/types within a functional group have different capacities to respond to different kinds of disturbances, and the range of different response types available is referred to as *response* diversity; it's this aspect of diversity that is critical to a system's resilience. It's akin to risk insurance and portfolio investment, something easily understood by anyone involved in business management. Where in the system is there only one way of carrying out a vital function? The shift to monospecies cropping is one example—and the decline in genetic diversity (varieties) within crops like wheat and rice is a worrying global trend.

We have dwelt at some length on diversity as an attribute of general resilience, but a similar questioning approach needs to be brought to bear on the other attributes we've listed, as well as anything not in that list that is revealed by the questioning process as potentially limiting to general resilience.

As a concluding comment, we'll remind you that efficiency-driven systems are likely to lose resilience. We raise this point again because "efficiency" has become a sort of undeniable goal for good business, government, and development. Cutting budgets via "efficiency measures" is standard practice in bureaucracies. However, where it removes response diversity, it comes at the expense of resilience (see box 4)

Openness

Openness refers to the ease with which things like people, ideas, and species can move into and out of your system. Closed communities of people and society can become inbred, static, and fragile. The same can happen with isolated patches of native vegetation.

As with diversity, there is no "optimal" degree of openness. Its effects depend on how resilient or nonresilient the system is in other ways, and either extreme (too open or too closed) can reduce resilience. What trends are occurring? Is there any evidence (social or ecological) that the system is becoming (or is) too closed?

Reserves

In general, more in reserve means greater resilience, and the trend to look for is often one of a loss of reserves—natural (e.g., habitat patches, seed banks), social (memory and local knowledge), and economic (levels of savings).

Can you identify any reserves that have come into play in the past, and are any of them changing? Examples include underground water supplies, reserve grazing areas and/or fodder banks, and people who know things about the area (corporate knowledge) that are irreplaceable. What changes are occurring, and are any trends worth flagging as something of concern?

Tightness of Feedbacks

As social-ecological systems develop, there is often a trend toward lengthening times for responses to signals, loosening the strength of feedback signals. It comes about through increasing complexity and levels of governance, increased steps in procedural requirements, and the weakening of the costs-and-benefits feedback loops involved in using resources. The signal of the environmental costs of a product from Africa to a buyer in a European supermarket, for example, is very weak.

All systems are kept in their current configuration (system regime) by critical feedbacks—environmental and socioeconomic—and weakening feedbacks reduces resilience.

Can you identify changes in any feedbacks (social, ecological, economic) that might be of concern? From your developing model of the system, can you identify critical feedbacks that act to keep the system in its current state—and are any of these changing, or weakening? In his book *Fragile Dominion*, Simon Levin (1999) gives examples of how privatization and changes in property rights strengthened feedback loops in several developing country regions. Have such changes influenced feedbacks and hence resilience in your system?

How about bureaucratic gridlock in which the amount of red tape around doing something gradually makes it harder to respond to environmental or social opportunities or stimuli, or to do anything? Sometimes the bureaucratic transaction costs grow to such an extent that a process comes to a grinding halt. This is symptomatic of a weakening feedback process due to increasing complexity.

Modularity

Again, there is no optimal degree of modularity, but a system that is fully connected will rapidly transmit all shocks (e.g., a disease, a wildfire, or a bad management practice) through the whole system. In a system with tightly interacting subcomponents that are loosely connected to each other (i.e., a modular system), parts of the system are able to reorganize in response to changes elsewhere in the system in time to avoid disaster. In highly connected systems, "the" successful way of doing things spreads quickly across the whole system. In modular ones, a variety of ways of doing things is maintained.

In what ways is the system modular? Are there any trends in this modularity? Is the system becoming more fully connected, or are there parts of it that are becoming more isolated, or too loosely connected? Do any of these warrant further investigation?

In the past year in Australia, two reportedly small computer glitches closed down an airline and a bank for several days. This is symptomatic of overconnectedness and a lack of modularity. In a modular system the fault would have been confined to a subcomponent, and the larger business could have continued to operate. The famous New York blackout of 2003, which actually affected most of the northeast United States, resulted from a lack of modularity in the power network.

These types of disturbances (like computer glitches) frequently take down just-in-time business operations where there are no reserves— where goods are couriered in rather than maintained as reserve supplies at the point of sale. So the interaction between attributes of general resilience can compound the consequences of their loss.

Leadership, Social Networks, and Trust (Social Capital)

These three, intertwined social attributes emerge repeatedly from case studies of resilience as important contributors to the "coping capacity" of a community. They are often referred to as "social capital." This term draws criticism from economists because social capital isn't a capital stock, but it is very important in conferring general resilience, and it's useful to consider the three main attributes that make it up. Without them the response capacity of the social-ecological system to disturbances is low. Social scientists talk about the two components of social capital being "bonding" and "bridging." Bonding is largely about trust but also leadership, and bridging is about the functionality of social networks.

The leadership attribute can be a tricky area for stakeholders to investigate, since it invariably involves the current leaders. However, it is helpful to consider leadership as a process, rather than as vested in one individual, and to recognize that different kinds (styles) of leadership are required for different circumstances. In good times, the appropriate leadership is of the "leading from behind" kind, while in times of crisis, strong individual leadership may be called for. Is the system locked into one style of leadership, or can it change to suit the circumstances?

Social networks are a major source of resilience, and it is important to recognize that there are often networks of networks—that is, different kinds of networks (business, religious, sporting) connected by common membership. During recent floods in Victoria in Australia, a rapid response to identify and assist the most vulnerable people was enabled by use of the network of Meals on Wheels (a charity organization that provides meals to society's most disadvantaged).

"Shadow networks" have emerged as being very valuable when a shock occurs. They are informal networks that can act quickly and effectively when needed. A shadow network is not active all the time and so does not have maintenance costs, but it can be quickly brought into play without the normal period of trust building required for a network to operate.

Social networks lead into the area of trust. It is a crucial component of community resilience, and though everyone has some idea of what it is, it is difficult to define or measure. An important feature about trust, however, is that it takes a long time to build but can be lost very quickly. As globalization and megacities grow, and regional connections in farming and other resource-use systems increase, the general resilience of these social-ecological systems will be determined by ways in which they can develop effective social networks that give rise to trust and rapid response.

Together with the kind of relationships a society has with the scales above (governance and level of support), these attributes determine the society's empowerment, its degree of "agency," which Brown and Westaway (2011) suggest is the capacity of individuals and groups to act in making their own choices. Having agency means that people are not just powerless victims of environmental and other changes. Assessing the status of social capital in a community requires looking at all of these as an intertwined dynamic process and exploring the changes that might be occurring in the three components of social capital—leadership, social networks, and trust.

Levels of the Capital Assets

A final category that confers general resilience is the amount and quality of capital assets the system can draw on in response to a disturbance. These include natural capital, built capital, human capital, and financial capital. Many of the desperate social-ecological systems in Africa's Sahel are trapped in very bad basins of attraction because they have low levels of human capital (education, health), very little built capital, and degraded natural capital and cannot draw on financial capital to help them change. Social-ecological systems differ in terms of which types of capital may or may not be limiting, and therefore which need to be addressed/enhanced in an effort to increase their general resilience.

The Essence of General Resilience

We say again, general resilience theory and practice require more development and research. What we have suggested amounts to some guidelines for approaching resilience, and you need to ensure you don't get bogged down in details. But, having made a list of some of the seemingly important things that are changing in terms of your system's diversity, openness, reserves, feedbacks, modularity, and levels of different types of capital, see if you can answer the following questions. Some of the answers will emphasize what you've already listed; some may have you adding new items.

- What has conferred "coping capacity" to your system in times of trouble? What worked in the past? If there were past failures, could they be attributed to any of the features conferring general resilience?
- Is there anything that is worrisome now?
- In a time of trouble, how good are the cross-scale connections and connections within the focal scale? Are there missing connections, especially between the focal scale you're interested in and scales above and below? When disaster has struck, were state and federal officials and politicians responsive? Was there a constructive community response (how good are the networks within your focal scale)?
- Are there any trends in any of the attributes in the list above?

Box 4: Resilience and Efficiency

"Efficiency" and cutting costs through "efficiency gains" has become a mantra of modern management. It is commonly held up as an important policy goal. But sometimes efforts to increase efficiency come at the cost of reducing resilience, and in some cases the net costs to the whole system can be high.

There is a cost to maintaining resilience, often in the form of forgone extra profit in the near term, and this gives rise to tensions with more conventional approaches to resource management—approaches that highlight increases in productivity based on narrowly defined efficiency. The tension comes down to comparing the costs of maintaining resilience versus the costs of not maintaining it. It's analogous to comparing the costs of insurance versus the costs of not insuring.

The comparison is fairly straightforward if you can estimate the probability of crossing a threshold and the probable costs associated with this crossing. Such calculations can be quite challenging to undertake, but they are worth attempting. A preliminary estimate of the two probabilities for a water table threshold leading to a salinized cropping area, for example, suggested that it was worth spending a lot of money to keep away from that threshold, even at high discount rates (Walker et al. 2009).

Efficiency is, of course, not necessarily bad, or anti-resilience. It reduces resilience when actions based on narrowly focused efficiency remove response diversity, and it's important to be clear on what we mean by the term *narrowly focused*. By this we mean not taking into account the secondary effects of increasing the efficiency of using some particular resource. Increasing the efficiency of nitrogen fertilizer use by crops, for example, does not have unintended secondary effects that then lead to loss of resilience. In fact, being able to reduce fertilizer applications in agriculture through more efficient uptake by plants is likely to increase resilience of the agricultural system, and avoid negative secondary effects such as water pollution. Likewise, increasing energy efficiency is a necessary development in combating climate change, but it needs to be considered in a holistic way.

Taking a holistic, systems view before deciding on efficiency actions allows you to distinguish between those likely to be resilience-negative and those likely to be resilience-positive (or neutral). Such consideration leads us back to the idea of interacting bundles of ecosystem services and the resilience of the delivery of the whole bundle of services. Unrecognized, or unaccounted, losses of ecosystem services due to pursuit of efficient use of just one ecosystem service reflect a decline in general resilience.

General Resilience, Adaptive Capacity, and Specified Resilience

While general resilience is not about preparing for specific thresholds, the likelihood that a system will cross a threshold depends on both specified and general resilience. The ability to keep away from a threshold in the first place, maintaining a large safe operating space, depends largely on the attributes that make up general resilience. They are much the same as those that have been described for adaptive capacity—the capacity to deal with a shock when it happens (like slowing the spread of a disease, or substituting one energy supply to an industry with another when the first one fails) so as to avoid crossing the threshold.

Transformability

Sometimes when you analyze the dynamics of a system, it becomes apparent that there's not much you can do to stop it from developing into something you don't want it to be. There are limits to how much you can adapt. If this is the case, then efforts to keep the system you now have won't help. Further investment is akin to digging the hole you're in deeper, and the first rule of holes is, when you're in one, stop digging. Now there's a need for transformation, to re-envisage what the system might become.

We deal with transformation in this chapter on assessing resilience because transformability, the capacity to effect transformational change, is part of the suite of capacities that add up to a system's resilience. In particular, in order for the system to be resilient at one scale, it is often necessary for some parts of the system at another scale to transform. Transformability is sometimes referred to as transformative capacity.

A current debate in Australia, for example, concerns transformational changes in its biggest agricultural heartland, the Murray-Darling Basin. It has become apparent that there is just not enough water in the system to maintain all the irrigation systems along its length (water was overallocated in early years when there was not enough knowledge about long-term water supply). So, in order for the Murray-Darling Basin to continue as an agricultural region, with its cities, towns, and rural populations, and in order for it to be resilient in the face of climate variability and climate change, it is necessary for some of the irrigation areas, or some parts of them, to undergo transformational change into some other kind of social-ecological system.

Transformability depends on three main attributes:

- Getting beyond the state of denial (acknowledging the need for transformational change): This involves raising awareness; the use of scenarios to explore possible futures has proved to be helpful in achieving this recognition.
- Creating options for transformational change: Though transforming the whole system at the focal scale may sometimes be possible, and therefore worthy of investigation, it is more likely that transformational change at the whole of the focal scale is too hard and too risky. It would therefore likely be deemed socially unacceptable. To move forward, the system needs the support and fostering of transformation experiments at finer scales that enable the exploration of novelty in "safe arenas." These experiments at finer scales need support from both the focal scale and higher scales (see figure 10).
- Having the capacity for transformative change: Transformative change needs support from higher scales and also depends on having high levels of all types of capital—natural, human, built, financial, and social.

Very importantly, the connections to scales above the focal scale need to be analyzed. Some of those connections might be good, in that they can help us transform, but some might be bad and hinder efforts. For instance, sometimes the existing "help" rules and arrangements are prescribed in ways that preclude the funding of novel activity. They provide, effectively, help *not* to change, as opposed to help *to* change. Subsidies for experimentation are clearly a good idea, whereas subsidies to keep doing what clearly isn't working, and one-size-fits-all regulations that prevent trying some new idea, don't help.

So, see if you can work out what connections are positive and which are negative, and then determine if, and how, they can be changed. Also, what other connections could there be that would help the transformation process (i.e., are there "missing" connections)? Can they be created?

The set of connections that together constitute transformability is depicted in figure 10. The ideas behind this diagram emerged from a workshop on transformational change in the Wakool Shire of New South Wales in Australia. The shire is at the lower end of the Murray River, and at the time of the workshop in 2010, irrigators there had

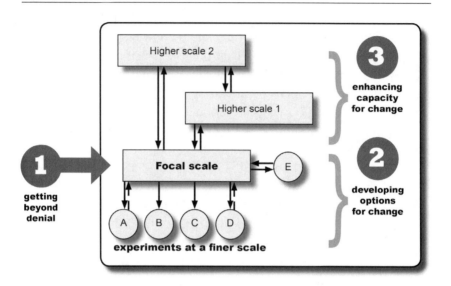

Figure 10: The Components of Transformability

Transformability requires (1) getting beyond a state of denial, (2) creating options for change at the focal scale, and (3) capacity for transformation, which relates to connections between the focal scale and higher scales. Creating options for change at the focal scale requires experimentation at a finer scale (A, B, C, D, and so on), though in some circumstances a trial at the whole focal scale may be appropriate (E). Many of these experiments won't work but some will, as indicated by the dashed arrows from A and D, feeding back to the focal scale.

received no water allocation for five years. The community had moved beyond the state of denial and were engaged in trying to identify new options. The question was how this process could be facilitated and enabled.

In our example from the Murray-Darling Basin, figure 10 means that help, in the form of financial and knowledge assistance, needs to come from the federal government scale, the relevant state governments, the Murray-Darling Basin Authority, and the relevant catchment management authorities. And this help needs to include appropriate changes to governance—the "rules" themselves, as well as who makes decisions at what level.

An equivalent set of governance scales and operations exists in all the regions/countries we have been involved with or have knowledge of, and all of them exhibit equivalent problems of incompatibility, arguments, and turf wars. It is often a major hindrance to both general resilience and transformability. It is here where the role of bridging organizations assumes great importance (see the discussion on social networks in the section on general resilience). It was just such an organization that broke a deadlock in resolving the conflicts over what to do about water in Sweden's Kristianstad "water kingdom" where declining water quality was threatening the natural and cultural values of the wetlands surrounding the city of Kristianstad (case study 5 in *Resilience Thinking*).

At the same time, innovations and experiments at the finest scales (individual farmers, fishers, communities of farmers, fishing villages, local businesses, and so forth) need to be encouraged as much as possible by those whose concern is the resilience of the focal scale. A delicate balance arises at this point. Any such program to foster novelty and experimentation calls for an assessment of each proposal in terms of its possible unintended, systemwide, secondary consequences. But this necessary process should not amount to more than putting a wider and longer perspective on the emerging proposals. It should not morph into a sieving process whereby a few people on a committee select what they think are good ideas.

Of the proposed experiments that are "safe," some, even if successful, may individually not amount to a transformational change. But perhaps a package of them might constitute the necessary critical mass that leads to a change in the way the whole focal scale functions. We repeat the warning that the "enhancing-capacity" process could become a bureaucratic sieving process, which would be counterproductive. The rule should be to allow all safe experiments to proceed, and "allowing" them may require some changes in governance at various scales.

And, just as there is an interplay between general and specified resilience, clearly there is interaction between these and transformability as well. If your system is stuck in a state of denial, has no options or capacity for change, then it has little transformability. The reason for this situation will relate to its general resilience. Strong general resilience (high levels of diversity, openness, connectedness, social capital, and so forth) almost by definition confers high transformability.

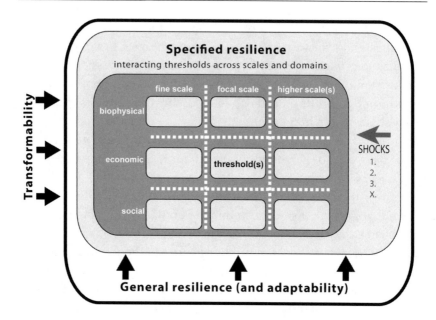

Figure 11: Specified Resilience, General
Resilience, and Transformability

These are different but interacting capacities of the system. Assessing a system's resilience requires an accounting of all three.

A Resilience Assessment

Using the descriptions of your system (chapter 2) and reflecting on how people believe that these various components are interacting, you have now

- Formed an agreed-on mental model of how parts of your system are working and where thresholds might lie (specified resilience)
- Reflected on your system's ability to cope and identified some attributes of the system that may be limiting its general resilience
- Considered your system's capacity to transform if needed (transformability)

That, in essence, is a resilience assessment and is summarized in figure 11. That this is a useful way to frame the assessment is supported by the fact that our colleague Paul Ryan independently came up with essentially the same picture through resilience workshops he has run.

Key Points for Resilience Practice

- Resilience is not a single number or a result. It's an emergent property that applies in different ways to the different scales, domains, and cycles (and their interplay) that make up your system. It's relative and contextual.
- Assessing resilience involves understanding specified resilience, general resilience, and transformability.
- You assess specified resilience by identifying alternate states and associated thresholds. This might be approached by considering known thresholds, thresholds of potential concern, conceptual models, and analytical models.
- Diversity, modularity, the tightness of feedbacks, openness, reserves, and high levels of all types of capital (including social capital) are important system attributes conferring general resilience.
- The attributes of general resilience interact. It is not possible to determine one particular level or amount of any attribute that marks a critical level. The most appropriate approach is to try to identify trends and changes and examine them in terms of possible effects.
- Transformability depends on three main attributes: getting beyond the state of denial, creating options for transformational change, and having the capacity for transformational change.

CASE STUDY 3

Assessing Resilience for "the Plan":

The Namoi and Central West Catchment Management Authorities

In many places around the world *resilience* is appearing in policy and mission statements. In New South Wales (NSW) in Australia, for example, the goal for natural resource management is "resilient, ecologically sustainable landscapes functioning effectively at all scales and supporting the environmental, economic, social and cultural values of communities" (NSW Natural Resources Commission, 2011).

That's quite a wish list, and it's not surprising that many policy makers and managers are somewhat daunted when it comes to turning such an aspiration into reality. What does *ecologically sustainable*, for example, actually mean? Put a group of scientists and managers in a room together to define it and you're guaranteed a lengthy debate, and that's a debate that's been firing for decades.

Resilience, by comparison, is the new kid on the block in terms of its addition to the lexicon of natural resource management, and while many precise definitions exist for what it means (this book is based on one set), there are also many approaches to the concept of resilience from different disciplines. Consider the discussion in chapter 5.

Beyond formal disciplinary definitions, we all have our own perceptions of what resilience is, a bit like how everyone has an idea of what health is, and for most people resilience is a good thing and the aim is simply to become more resilient. As we've already described, if you apply our definition of ecological resilience, being more resilient is not necessarily a good thing. If you're stuck in a bad place, it's undesirable, but the normative view (that more is better) needs to be kept in

Image 5

Multiple land uses in the Namoi catchment (sunflowers, oats, cattle, remnant native vegetation). (Photo: Namoi CMA.)

mind when catchment managers are charged with creating "resilient, ecologically sustainable landscapes," because their stakeholders bring to the table a range of ideas on what resilience is.

So how does this goal of managing for "resilient landscapes" play out in real life? In NSW two groups of catchment managers are attempting to apply resilience thinking to their own planning processes.

NSW is divided into thirteen major catchment areas, each with its own catchment management authority (CMA) that is responsible for developing a rolling catchment action plan (CAP) for achieving the catchment's goals. The NSW Natural Resources Commission, charged with taking a whole-of-government perspective, oversees this process. In 2009 the commission adopted a resilience approach for developing the CAPs and assisted two CMAs, the Namoi and the Central West, in running pilot resilience assessments to see what value the approach might hold. The resilience assessments have been used in developing revised pilot CAPs by the two catchments. They are probably the first examples of regional strategic plans based on resilience assessments (NRC 2011).

Image 6

Stakeholders in the Central West catchment wading through the internationally re-nowned Macquarie Marshes during a resilience workshop. (Photo: B. Walker.)

Neighboring Catchments

Lying in northwest NSW, the Namoi catchment covers some forty-two thousand square kilometers, mostly along the Namoi River and its tributaries. It is home to around a hundred thousand people, with major industries including cotton, livestock production, grain and hay, poultry, and horticulture (see image 5). Irrigated agriculture is one of the defining characteristics of the region, and it is heavily dependent on groundwater. The region also contains the largest continuous remnant of semiarid woodland in temperate NSW, known as the Pilliga Scrub, and declining biodiversity is one of the major challenges faced by managers. (For more information on the Namoi see http://www.namoi.cma.nsw.gov.au/.)

Immediately to the south of the Namoi catchment is the Central West catchment, which includes the Castlereagh, Bogan, and Macquarie River valleys. Covering eighty-five thousand square kilometers, it's twice the size of its Namoi neighbor but not as dependent on irrigated agriculture. Agricultural land dominates the Central West, generating over AUD$1 billion annually from cropping and grazing. The region includes the internationally recognized Macquarie Marshes, a Ramsar-listed wetland important for bird breeding (see image 6). (For more information on the catchment see http://ow.cma.nsw.gov.au/.)

There's a lot in common between the Namoi and the Central West: they operate within a common framework of law, both are agriculturally based, and both share issues on declining biodiversity and environmental degradation with a scarcity of available resources to deal with the challenges. But they are also quite different, with industry in the Namoi being more dependent on irrigation, in particular from groundwater, and with increasing interest by the mining industry in the coal deposits and coal seam gas—and mining uses a lot of water. The development of these industries is creating tensions over water use.

Not a New Start

As will be the case with any group wishing to adopt a resilience framework, applying a resilience approach to planning and management does not mean starting from scratch. Each catchment already has a large body of work, with many existing government reporting requirements that have had to be complied with.

Each catchment has its own CAP that needs to address multiple plans and policies, for example, water-sharing plans (gazetted under the NSW Government), a biodiversity strategy (drawn up by the NSW Government), and a Murray-Darling Basin plan (created at the Australian Government level, in association with state governments, that sets out water allocations across the broader basin of which these catchments are a part).

Adopting a resilience approach, therefore, involves putting a resilience lens over the existing strategic and operational plans to see how and where such a perspective suggests things need to be changed, and what new things need to be included.

A Format for a Regional Resilience Assessment

A resilience assessment isn't achieved in a day. It takes a concerted effort over weeks and sometimes months, with multiple exposures to the concepts underpinning resilience thinking, and time for reflection and honest discussion. And, as both catchment groups soon realized, for many of the people who need to be engaged, time is usually in short supply. So lesson one was that undertaking a resilience assessment shouldn't be done lightly. If you are going to attempt it, acknowledge up front that it is going to require time and money to bring stakeholders together to make it work.

The format that emerged in the Namoi and the Central West involved an initial two-day session that covered three main topics. This first session included key stakeholder representatives, many of whom were members of the CMA Board (appointed on the basis of representation and expertise). It was followed by a series of three or four other sessions (each taking two to three days) on particular aspects of the catchment's resilience. The initial session covered three components (largely in alignment with the description outlined in chapters 2 and 3):

1. Group Discussion on "What Is Your System?"

The sessions began with a two-hour presentation/discussion on resilience, with stakeholder questions throughout. This was necessary to get everyone "in a resilience frame of mind"—being clear about what resilience is and isn't, and understanding the main concepts. They then tried to answer the following questions:

- What is the system? (What are the component parts—ecosystem types, land uses, rivers, towns, etc.?)
- What is the "focal" scale, and what are the important scales above and within it?
- What do you value in, and from, the system? (What is it you wish to make resilient?)
- What are the big issues, or problems?
- What are the drivers of change, and what trends are occurring?

2. Group Session on Assessing Resilience

After describing the system, the next step is to attempt to assess its resilience, and this involves considering the three aspects of resilience:

- Specified resilience: attempting to identify at different scales and in each domain (biophysical/ecological, economic, social) known, likely, or potential thresholds, with their controlling variables and drivers
- General resilience
- Any needs for transformational change, and the level of transformative capacity

Note the point we made at the beginning of chapter 3: the order in which you do these is not as important as the need for iterating between them.

As discussed in chapter 3, trying to identify thresholds is not easy. But the pragmatic way the CMAs approached it was to start by identifying known or strongly suspected thresholds, the "thresholds of potential concern."

Beyond what was immediately known or suspected, the next level of assessment was to develop conceptual state-and-transition (S&T) models for key parts of the system or the system as a whole. It involved describing the current state of the system or subsystem, then what other possible states it could be in, and the necessary conditions for transitioning from one state to another. The transitions were then examined for how reversible they were and whether thresholds might be involved.

The development of such S&T models could not be achieved in the first two-day sessions with these CMAs, but the attempt constituted a product that was dealt with in subsequent, smaller group sessions. The next step would be the development of more detailed analytical models to explore the nature and positions of thresholds. This work has been done in Central West, using detailed cause-and-effect modeling of each process operating within the S&T models. These models examine the effect of modifiers operating with a system that may cause changes around a threshold.

There is no ideal format for this analysis of threshold effects, and no "best" framework, but the framework that proved useful in these two cases was of scales and domains, as shown in figures 6 and 7.

3. "So What?"

What are the options, and some proposed actions, for intervening in the system to manage resilience?

- Possible and appropriate interventions
- Kinds and scales of interventions
- Sequencing of interventions
- Developing an outline of an adaptive management program (treating policy and management as an experiment)
- Where transformation is called for

Through a Resilience Lens

The Central West group quickly saw the value in creating conceptual S&T models for the various subcomponents of their catchment and also applied this approach to the social and community aspects of their region. All known evidenced-based thresholds were identified: for example, thresholds of vegetation cover at 70 percent and 30 percent; vegetation patch size, shape, and proximity; species population sizes; groundcover at 70 percent; and soil carbon ranging from 1.0 to 1.6 percent by weight depending on altitude. They also indentified specific thresholds for farm viability by examining debt, income, and equity ratios. Targets were developed around staying within the limits of the system to avoid crossing undesirable thresholds.

The Namoi attempted to identify critical thresholds in relation to the targets they have to report on. Few could be quantitatively determined, and some fell more into the category of desired (utility) threshold levels, as opposed to transition points between alternate stability regimes. Nevertheless, these have proved useful. For example, in regard to the issue of "soil health" (the problem is soil erosion and loss of fertility), they identified 70 percent groundcover of vegetation as a critical threshold level and then proposed actions for staying above this.

With regard to the declining state of riverine ecosystems, the Namoi CMA identified a "safe" threshold level of 66 percent of predevelopment surface water flows, and they proposed actions for staying above this level. This is an appropriate starting point, and more work will allow refinement of the actual level in their region.

Adding Value

Unlike other management approaches that seek to measure and improve on the condition of specific environmental or economic assets, a resilience assessment doesn't produce some particular level or amount of resilience that can be compared between catchments. The Namoi was not (and probably could not be) shown to be more or less resilient than the Central West.

What was achieved was an engagement in the catchment's complexity by most of its important stakeholder groups. These people—managers, farmers, businesspeople, and so forth—were encouraged to view their catchment as a self-organizing system that changes over time, but with limits to this self-organizing ability, and hence a system that could lose its identity if it moved too far in certain directions.

In the Central West the assessment encouraged an attempt to conceptualize how the system could exist in a variety of states, and the pathways between them. The stakeholders were attempting to understand what was giving their region its identity, and how this could change and what might cause that change.

In the Namoi they attempted to identify critical levels of change beyond which their system might organize into something different. They also made clear connections between the condition of their groundwater resources and the level of cover of natural vegetation. They acknowledged the connection between the groundwater resources and the economic and social well-being of the region.

Both catchment action plans are now completed for this stage. Both CMAs agreed that the resilience assessment was worth the effort. The Namoi reported that the resilience framework provided "a fresh lens to look at tired problems." Their management board felt that because of working through the resilience assessment, there had been an "uncluttering of the NRM agenda." Central West CMA reported that the process gave them a more focused approach and an ability to examine how each system may impact on another.

In an assessment of the value of applying a resilience framework to catchment action plans (NRC 2011), the Natural Resources Commission said the Central West and Namoi CMAs showed that resilience thinking

- Helps develop a holistic picture of how the landscape functions and test assumptions
- Helps manage complexity by focusing on the few most important things
- Is a useful concept to engage the community in strategic planning
- Embraces change and builds capacity to manage for natural variability and extreme events.

Each CMA has developed a report of its initial assessments. These can be accessed online: at http://www.namoi.cma.nsw.gov.au/274456.html for the Namoi plan and at http://cw.cma.nsw.gov.au/AboutUs/strategicplanning.html for the Central West plan.

A Resilience Lens on the Macquarie Marshes

The Central West CMA has confirmed its adoption of a resilience approach to planning and development and recently conducted a follow-up resilience assessment of one part of the catchment, the Macquarie Marshes. Containing Ramsar bird-breeding sites, the marshes are a particularly valuable, high-profile, and contested region. They include some state-owned nature reserves, but most of the marshes are privately owned and the floodplain areas have been used for over a hundred years by livestock farmers. Irrigation farmers adjoin the region, and the flows of water have been regulated for decades, controlled since the middle of the last century by the large Burrendong Dam some 250 kilometers upstream. Water allocations to irrigation and nature have led to ongoing conflicts between the three stakeholder groups (nature, graziers, and irrigators).

The assessment took the marshes as the "focal" scale, the surrounding regions and the CMA area as the scale above. The embedded scales were the "core" marsh areas, floodplains and other ecosystems, and the different land uses. The initial two-day session identified some nine known or potential water-related thresholds affecting each stakeholder group, at different scales and in different domains, including critical threshold levels of flooding for persistence of the core marsh ecosystems.

It also identified some interventions that could be made to manage some of these thresholds. In chapter 3 we discussed one of them, a critical dispersal distance for native fishes, as an example of an embedded threshold. If dispersal distance is limited by dams, levees, channels, and so on to below some critical level, native fish species don't persist, though the introduced carp do. The introduction of fishways was identified as an intervention to increase the dispersal area and hence fish resilience. Another intervention that was considered involved changing the size of intake valves in the main dam above the marsh to allow for more flexible water-flow volumes, and hence the sizes of floods over the floodplains.

The work continues, with the aim of developing an adaptive management program to introduce interventions for managing the resilience of the whole "bundle" of goods and services coming out of the Macquarie Marshes. An initial outcome (expressed by a senior landholder) was that, for the first time, it seemed like addressing the interactions between these valued ecosystem goods and services in an agreed-on assessment framework had "got everyone on the bus."

Resilience Practice and Managing Agricultural Catchments

A resilience assessment doesn't happen in a vacuum; every region already has a set of plans and strategies in place, and these should be used to inform the resilience assessment. A resilience assessment also doesn't occur over a lunch break. If an assessment is to be done, it should be properly resourced and done in an iterative way. It is important to allow time between intensive sessions for reflection and follow-up activities. A resilience assessment enables key stakeholders to consider their system from new perspectives.

4

Managing Resilience

iven what you've learned about the system, given your assessment of real or suspected thresholds, the system's general coping ability, and capacity for transformation—so what? What should you do about it, and what options are available to meet these concerns?

A resilience approach to management involves the development of an adaptive management and an adaptive policy (governance) program in which the interventions are considered experiments that test the assumptions that gave rise to them. Don't just start with one or two immediate interventions that seem most obvious. It is important to consider the full set of possible interventions and to develop them into a logical sequenced program.

Some of the interventions will be about dealing with specific thresholds. Some will be about attempting to initiate necessary transformations. Others will be about changing how decisions are made and how management handles uncertainty (general resilience).

Tools and Options for Management

Where, when, and how to intervene in a social-ecological system ultimately comes down to the tools and options you have available. The assessment of resilience will have resulted in three main kinds of information:

- Specified resilience: Some version of a scales × domains thresholds matrix, akin to those in figures 6 and 7, should identify a priority set of known and suspected thresholds. Depending on how much work has been done, the information on the thresholds can vary from stakeholders just being aware that they exist, or might exist, through to a detailed knowledge of the attributes of the system that determine where on the relevant controlling variables the thresholds lie, and how far the system is from the threshold.

- General resilience: You should have knowledge of (or at least some initial ideas about) which aspects of the system are of concern in terms of reflecting low levels of general resilience and the adaptive capacity of the system.
- Transformability: You should know something about the capacity and readiness of the system to undertake transformational change, if needed.

The questions to ask now, considering all three kinds of information, are, What kinds of interventions are called for? What actions would be most appropriate, how could they be applied, and at what scales? It is helpful to consider these questions under four main kinds of interventions:

- Management (changes in recommended management)
- Financial intervention (assistance, investments, subsidies, taxes)
- Governance/institutions (including laws and regulations, policies)
- Education/information (to influence behavior)

Each of these categories of intervention operates over a different time scale. Management actions might be needed immediately to prevent a system from crossing a threshold, whereas education might influence behavior over decades but create a fertile environment for improvements in governance. They are complementary and should be considered together as a package. And this consideration should also include how best to sequence actions.

A common problem arising when interventions are recommended from a disciplinary viewpoint (e.g., from an economic, agricultural, ecological, or social point of view) is that each discipline sees the priority lying within its own realm. Economists usually come up with economic interventions, ecologists will recommend changes in ecological management, and so on.

This is where the value of a multi-stakeholder assessment group comes to the fore. Explicitly asking a mixed group to consider each of the four kinds of interventions overcomes the "if all you have is a hammer, everything looks like a nail" problem. The group as a whole needs to consider which kinds of interventions are called for, which will be most appropriate, and what the sequence for implementing them should be.

Also keep in mind that existing strategic plans and operational plans will already usually exist in the region being considered. Consider, for example, the various plans that the catchment management authorities needed to comply with in case study 3. Most of these plans will contain important and useful information. And, further, much of what these plans recommend is justified and correct. However, if they're driven by production imperatives and have been developed with an underlying philosophy that assumes smooth responses to drivers of change and that does not consider secondary effects and their feedbacks, they may well have some fundamental flaws that could result in unwanted and unexpected surprises.

The first step, therefore, is to place a resilience lens over the existing plans and note how they would be different had they been developed with resilience in mind. It may well be that this exercise is all that is required to "put resilience into practice." There will be other cases, however, that call for more radical changes, and it is therefore useful to consider how one might go about developing a program for resilience management.

Adaptive Cycles: When and Where to Intervene

Before intervening you also need to consider where your focal scale sits in an adaptive cycle and what's happening at the scale(s) above and below.

You'll remember from chapter 1 that the system is always changing as it moves through time, because of internal (endogenous) processes. It's moving through a cycle in which the connections between the components that make it up are weak, become strong, and eventually break apart. This is the adaptive cycle. Sometimes things are in gridlock, sometimes they're in freefall, but most of the time they're somewhere in between. It'll be easier to make changes at some times than at others. Sometimes the system will be brittle to small shocks, whereas other times it will be resilient to them.

And adaptive cycles occur at all scales, so just as the system at your focal scale will be somewhere within an adaptive cycle, so will the system at the scale above and the scale below. The thing you're interested in, be it a farm, a catchment, or a business, is a dynamic system at one scale in a nested set of scales—a panarchy.

The nature of the connections between scales depicted in figure 10 on transformability, especially those to the scales above, is in large part determined by the phases of the cycle those scales are in. If the system at the regional government scale is in a vibrant development phase, for example, the influence from it on the (lower) focal scale, in terms of helping to change, is likely to be positive. Using that connection is therefore to be encouraged. But if this higher scale is in a phase of lockdown or upheaval, the influence is likely negative and the appropriate action at the focal scale is either to defer action until the connection is positive or to find ways to bypass the connection and avoid negative influences.

A system that is in a late K (conservation) phase, where things are in gridlock, is the most difficult to change. In this situation it may be most effective to focus on education, making all stakeholders aware of the resilience situation and its likely consequences.

This is often a time when there are strong calls for subsidies to continue with "business as usual" by those in dominant positions. Yet, it is the very time when assistance should be in the form of helping change, rather than resisting change. For financial interventions in systems in the K phase, this should be considered carefully. Unless there is willingness to consider it, other interventions may not be effective.

The K phase is, however, an important time for developing action plans for what to do when the gridlock is ended by some crisis. In the absence of such preparedness, opportunities can be lost and the ensuing back-loop phases can be long and expensive.

Systems that have gone into release in the early back loop (release phase) are chaotic and releasing capital of various kinds. This system is unlikely to be responsive to any particular recommendations of change. The best thing to do is to assist (enable) the system to move quickly into a positive, reorganization phase, where it is open to suggestions and seeking solutions. Changes in governance, and new policy suggestions, will be more likely to succeed here than if tried in the K phase.

In the early growth (r) phase, a system is still open to changes but it is a time for consolidation of what occurred while it was reorganizing and for making rapid progress and growth. It is a time for financial help, new policy development, and changes in management practices.

Sequencing of Interventions

Timing of interventions is important, and so is the sequence in which they are attempted. If resilience management calls for stopping the clearing of vegetation, for example, it is no good offering financial assistance without first changing the relevant regulations.

In his book *Globalisation and Its Discontents*, Nobel economist Joseph Stiglitz (2002) makes the point that it was mistakes in sequencing, forced by the International Monetary Fund, that resulted in many of the economic failures in Africa and Russia and the Southeast Asian financial crisis of the 1990s. Forcing liberalization before safety nets were put in place, before there was an adequate regulatory framework, was inappropriate. Forcing open-market policies that led to job destruction before the essentials for job creation were in place created enduring problems. And forcing privatization before there were adequate competition and regulatory frameworks led to many undesirable outcomes. All these interventions ended up having negative impacts rather than the positive outcomes that were intended.

The four kinds of interventions themselves constitute a system of interventions, and it is necessary to consider the interactions among them to determine the appropriate sequencing and pacing of whatever interventions are proposed.

Timing and levels of interventions of humanitarian aid were found to be crucial in promoting resilience in people in displacement camps in postwar Eritrea (Almedom et al. 2007; see chapter 5). Exactly what type of aid was needed depended very much on what those affected said themselves, rather than what others thought they needed.

Is Transformation Called For?

Some of the most challenging interventions will involve transformation—changing components and sometimes the scale of the system, or parts of the system. When done in a deliberate, positive way, transformation means developing a different way of making a living, and achieving this depends on having the capacity to do it—having the necessary amount of transformational capacity.

Transformational changes happen all the time, but they are usually unplanned and often involve unpleasant effects for those caught up in them. At the scale of a nation, Russia has undergone transformational change twice in the last century, from the system of czars to communism and then from communism to capitalism. Both times the transformational process had (at least initially) significant, adverse impacts on human well-being. At the time of writing, the Middle East is going through transformational change, and the costs are severe.

At the scale of an industry, overfishing on the Grand Banks transformed the Newfoundland cod-fishing industry, an industry involving many boats and people, to a lucrative long-claw crabbing industry involving only a few players.

At the scale of an international lake system, the Aral Sea was transformed from being one of the four largest lakes in the world to a baking plain with a once-prosperous fishing industry virtually wiped out—not just a change in state of the system but a change in the nature of the system (see box 5).

Changing technologies and markets have transformed vast swathes of once-prosperous manufacturing regions in the United States into rust belts, so called because when the region's factories were closed, the resulting shuttered buildings were guarded only by rusting gates.

Transformations, however, don't have to result in loss and destroyed livelihoods. In Detroit, once a giant auto industry manufacturer but now part of the rust belt, plans are under way to buy up great chunks of abandoned land to build what is hoped to be the world's biggest urban farm. All up, there could be some thirty thousand acres of land bulldozed and transformed into an urban farmland. Deliberate, positive transformation certainly has costs, at least initially, but they are likely to be far less than the costs associated with an inevitable, unplanned transformation that is avoided until it is too late.

Initiating transformational change in order to avoid having it done to you is still rather a new concept in the world of resilience science, and the determinants of transformability are not well-known. However, as described in chapter 3, they fall into three main classes: getting beyond denial, creating options, and having the capacity to change. It's worth making a few additional comments here in terms of their management.

A state of denial is common in regions facing difficult circumstances. You hear comments such as "we can deal with this," "it'll come right," "there's no need to change, we just need to get a bit more efficient," and so forth. Increasing efforts are made to keep going and

Text continued on page 126

Box 5: A Tale of Two Transformations

Here are two stories on regions that have experienced transformation in the past half century. One was deliberate with positive outcomes (initially anyway). One was unintended with catastrophic outcomes. Both underline the point that transformations usually come with big consequences.

Southeast Zimbabwe

For many decades cattle ranching in southeast Zimbabwe had been the dominant land use. But declining terms of trade (dynamics in the economic domain) and increasing amounts of woody shrubs (dynamics in the biophysical domain) had made it less and less profitable. Much of the region had shifted from a regime of productive, grassy savanna to one of low-production, shrubby savanna.

In the early 1980s there was a severe, two-year drought that led to 90 percent of the cattle dying. However, ranchers noted that the remaining wildlife (that they had been busy eliminating in previous years) fared much better. Rather than persist with an enterprise that was firmly lodged in what had become an undesirable system regime, many landholders reinvented their enterprise (Cumming 2005). They joined their properties, removed internal fences, and transformed their farms into game-hunting and safari parks. Their efforts met with enormous success, though subsequent political events at the national scale have once again created undesirable outcomes.

The story is one of declining resilience of cattle ranches in the rangelands, due to both ecosystem change (no browsing wildlife to control woody shrubs, coupled with prolonged grazing pressure) and declining terms of trade for livestock. This led to a regional-scale gridlocked K phase followed by a release induced by a major drought. And this led to a transformative change from grazing cattle to a landscape dominated by wildlife conservancies.

The Aral Sea

One of the best, though most tragic, stories of forced transformation can be found in the Aral Sea. Only fifty years ago the region boasted a stable and productive fishery with mixed agriculture in its deltas. Today it's an ecological basket case. (The following account is based on Micklin 2007 and Schluter and Herrfahrdt-Pahle 2011).

The Aral Sea is a terminal lake located amid the great deserts of Central Asia. In the two centuries prior to 1960 the lake had been a relatively stable system with levels fluctuating little more than four

Continued on page 124

Box 5 continued from page 123

meters. Instrumental measurements began in 1911 and recorded that up until 1960 the sea's water balance was remarkably stable, with average annual inflows of fifty-six cubic kilometers of water, roughly matching net evaporation.

With a surface area of slightly more than sixty-seven thousand square kilometers, the Aral Sea was the world's fourth largest inland body of water, a vast brackish lake inhabited mainly by freshwater fish species. And, despite being located in a region of deserts, it was a center of biological diversity and economic prosperity. The sea supported a major fishery and functioned as a key regional transportation route. The extensive deltas of the Syr Dar'ya and Amu Dar'ya sustained a broad diversity of flora and fauna, and the region also supported irrigated agriculture, animal husbandry, hunting and trapping, fishing, and harvesting of reeds.

But from the early 1960s things changed dramatically as expanding irrigation sucked water from the sea's two inflowing rivers. Irrigation of itself was not the death of a productive Aral. Irrigated farming had been practiced in the basin for millennia. However, until the 1960s most of it had occurred at a moderate scale in the deltas and edges surrounding the lake, meaning much of the irrigation runoff flowed into the basin.

In the 1960s, under the direction of the Soviet Empire (control from a higher scale), irrigated cotton production was massively extended in order that the USSR might gain independence from cotton imports. These new huge irrigation systems extended into the surrounding deserts, meaning the water taken from the rivers was now being lost to evaporation. During the 1960s the lake experienced a significant deficit in its water balance, around twelve cubic kilometers of water every year (i.e., water evaporated but was not replaced by inflow). Worse was to follow.

Water management was of a very low standard, with a massive overuse of water in agriculture and improper drainage. This led to waterlogging, soil salinization, and desertification of the deltaic wetlands and a significant reduction in productivity along with a devastating impact on biodiversity. To counter the declines in agricultural production, irrigation was pushed farther into the desert and massive drainage systems (technical solutions) were built to enable the leaching of soils to flush out the salt prior to irrigation. The strategy had an ever-mounting environmental impact.

The basin's water deficit grew dramatically to thirty cubic kilometers of water per year in the 1970s and 1980s, with many years seeing no inflow from the Amu and Syr. The Aral separated into two smaller lakes in the late 1980s (the Small and the Large Aral Seas). By 2006 the level of the small sea had dropped by thirteen meters, the large by twenty-three meters. Taken together the area of the two seas has decreased by 74 percent, the volume by 90 percent.

The once-prosperous fishing and fish-canning industries that had been processing between fifty thousand and three hundred thousand tons of fish per year went into terminal decline. They collapsed in 1982, throwing tens of thousands of people out of work. Navigation on the Aral also ceased in the 1980s.

The sea's shrinkage also led to climate change around the basin in a zone up to one hundred kilometers wide along the former shoreline in Kazakhstan and Uzbekistan. Summers have warmed and winters have cooled, humidity is lower, and the growing season is shorter. People living around the sea (in what some call an ecological disaster zone) suffer acute health problems. These relate to increased levels of dust and salt being blown from the desiccated basin, poorer diets due to the loss of Aral fish, and exposure to environmental pollution from the heavy use of toxic chemicals in irrigated agriculture.

The catalog of ecological and social consequences of the Aral transformation is too long to go into here. Many species and ecosystems have gone extinct or are on a fast track to extinction, agricultural productivity has slumped, and the region's resilience to shocks like droughts is critically low.

The prospect of a return to a thriving Aral Sea is remote, despite significant investment in returning some flows to the basin. The effort is too expensive, plus the system is locked into using whatever water is available to leach soils of salt to continue their dependence on irrigated agricultural production.

Schluter and Herrfahrdt-Pahlc (2011) believe that the Amu river basin, as an example of the challenges facing the broader region, is locked into this degraded condition by ecological dynamics and vested interests. To break out of it will require another and very difficult transformation involving changes at higher and lower scales in the system. Part of that transformation will involve moving away from relying solely on irrigated agriculture as the backbone of the economy and moving toward developing a diverse set of economic activities.

All the inland lakes of the world are under increasing pressure and have declining resilience, and another disaster, of equal magnitude to the Aral Sea, is now unfolding in Africa—the decline of Lake Chad. Originally an inland sea of twenty-five-thousand square kilometers supplying water and huge fish yields to Nigeria, Chad, Cameroon, and Niger, it has declined by about 90 percent because of a combination of changing rainfall and extraction for irrigation. About 50 percent of the decline is due to irrigation (Coe and Foley 2001).

to lobby support from governments to keep on doing the same thing; and governments all too frequently accede, taking a short-term view. Information is the main key to breaking a state of denial. Scenario development in which the range of possible futures is examined and spelled out is one good way of engaging stakeholders in the process. Peterson and colleagues (2003) describe how this might be approached and provide a number of examples from business, government, and conservation planning.

Identifying options for transformation mostly requires encouraging and investing in novelty and experimentation as well as help *to* change, rather than subsidies that allow people to continue doing the same thing when it is no longer viable. This flows on to the capacity to change, which depends on effective connections (for support) to the scales above, and high levels of all the types of capital. A major constraint in developing world regions faced with the need for transformation is that they are low in all types of capital.

So how do you proceed? Assuming that the very important first step—getting beyond the state of denial—has been achieved, the process of transformational change involves combining development of the options for change with support for change. Since transforming the whole focal scale is risky, expensive, and unlikely to be supported, what is needed is support for novelty at finer scales—an interactive process of support for experimentation between the scale at which transformation is needed and the scales below.

The Dutch have developed an approach to what they call "transitioning" in which they define "safe arenas" for experiments (see the website for DRIFT—the Dutch Research Institute for Transitions, if you want more information). They emphasize the importance of "niches" and "strategic niche management" in which such experiments are enabled. As some of them begin to show promise, they feed back up to the focal scale and so influence the directions of further support.

Transformation is best undertaken in a bottom-up manner; it is safer, offers more variety and opportunities to test different ideas, and is more likely to succeed. However, it is also true that in some instances it may not be possible to initiate the most appropriate changes without a prior transformation at a higher scale. Many desirable changes in energy and economic practices/policies at lower scales, for example, will not be possible without agreed-to changes at national and international levels.

From some of our recent work it is apparent that some parts of a region, some sectors, may need to be transformed in order for the focal scale as a whole to remain viable, while for other parts the appropriate thing is to maintain the resilience of the current system regime.

This leads to what is an increasingly important and common question that needs to be asked in all regions, and all countries: In which parts/sectors/industries/enterprises should we be trying to enhance resilience because they are in states that we like, that are good for us, and that have good future prospects, and in which parts should we be reducing resilience in order to ease transformation into a different kind of system? Such transformational changes increase the resilience of the larger system as a whole.

One final and very important point about transformational change: there is a danger in thinking about transformational change as a one-off thing, a struggle and cost that needs to be borne only once. It may be so in some cases, but the rates of environmental and social change in the world today suggest that we need to get into a mind-set of more or less continual transformational change. We need to consider it more like changing trajectories when necessary in order to avoid being trapped on a bad trajectory from which escape becomes increasingly unlikely, learning how to keep changing among a range of acceptable trajectories (systems with different identities) while avoiding those that are, or that have become, unacceptable.

The three components of transformability are difficult enough to orchestrate in a fluctuating but nontrending environment (social and natural). The world today is not like that. The speed and magnitude of directional change, environmental and social, are now such that what is needed is continual transformational change in human-dominated systems. The recent paper by Stafford Smith and colleagues (2011) makes the point strongly that adapting to a particular "new" climate is illogical, and that continuous change is needed.

Adaptive Management

The ideas of adaptive management arose in conjunction with the ideas behind resilience thinking (Holling 1978; Walters 1986), and they are an integral part of a resilience approach. Developing a program of interventions to address the resilience problems the assessment has revealed is best done within an adaptive-management framework.

The essence of adaptive management is treating management as an experiment, or to be more precise, treating it as a hypothesis coupled to a management experiment to test it. There are many references to "learning by doing" in discussions on adaptive management, but in practice, this is quite different. Learning by doing is usually carried out as a blind and noninstructive form of trial and error. It doesn't involve a prior prediction of the outcome of the "doing." Because of this, a trial-and-error approach very seldom helps us to learn. Indeed, despite its name, it's more about doing than learning.

Adaptive management requires, at the start, an explicit statement of the expected response of the system to some particular management or policy intervention. If, after the intervention, the system's response differs from this expected response, then your understanding of how the system works was wrong. You therefore need to adjust your model of the system accordingly, even if it is only a conceptual, mental model. Extending this, adaptive management involves the development of an evolving model of system structure and function as management and policy development proceed.

Adaptive management in its simplest form is known as "passive adaptive management," in which the model is adjusted and developed using whatever management actions are being implemented. "Active adaptive management" is a step further. It involves deliberately doing things to the system to learn about it, even though the intervention may not be in the immediate best interests of whoever is using the system. Experimenting to learn may be at the cost of short-term profits.

In a resilience context, it may be possible to use existing management results (passive adaptive management) to make inferences about where a threshold might be—especially if it has been crossed in some places. But it may also be that exploratory management interventions are needed to probe for the locations of thresholds.

In theory this all sounds great. In practice, adaptive management, and especially active adaptive management, can be challenging to implement. Carl Walters is one of the main architects and champions of adaptive management, and his analysis of why it often fails (Walters 1997) provides a good discussion on the challenges.

Policies of experimentation are often seen as too costly or risky to implement, especially when it comes to managing sensitive (politically and socially) species. And because adaptive management often shifts the status quo, stakeholders in research and management often resist it because adaptive-policy development can be perceived as a threat to existing research programs and management regimes. But this does not detract from its importance. An example of the difficulty of implementing adaptive management is discussed in chapter 5 in relation to managing the Everglades.

While we can make some good advances using passive adaptive management coupled with an explicit model of system response (conceptual at first, then quantitative), the use of active adaptive management deserves more consideration than it is currently given. Indeed, active adaptive management is an important part of putting resilience thinking into practice and should be explored as part of the framework for implementing the possible set of interventions that arise out of a resilience assessment. Establishing whether a threshold of potential concern is in fact a real threshold, and determining where on a controlling variable it might lie, will sometimes require experimenting with the system as part of management. And this should be as part of an ongoing learning program. This is what adaptive management is all about.

Beyond using it to explore particular thresholds, adaptive management needs to be an integral part of any policy development that embraces uncertainty and adopts resilience thinking. Developing a program of interventions to address resilience problems is best done within an adaptive-management framework. In chapter 3 we introduced the idea of structured decision making and the use of TPCs (thresholds of potential concern). The inclusion of both biophysical and "utility" (preference) thresholds, as in the TPCs and the structured-decision-making approaches, calls for an adaptive-management program, and the process described as "strategic adaptive management" (Kingsford et al. 2011) is a useful framework.

Strategic adaptive management (SAM) starts by defining the management objectives, in a hierarchical way, and this again highlights the need for the whole process to be iterative, because having this step in mind helps give some focus to the initial steps of describing the system: What it is that people value in and from their system and, therefore, the desired/undesired states of the system at various scales?

SAM then goes on to determine the management options for how to deal with the TPCs that have been identified and for how to operationalize them. It's here that the proposed management actions are built into an adaptive-management framework, using explicit predictions of the outcomes of the actions. The final step is the evaluation and learning part that feeds back to the evolving "model" of the system—that is, which predictions were wrong, and how does the model need to be changed so that it is consistent with the outcomes of the action?

The structured-decision-making framework suggests more precise, analytical models for including both biophysical thresholds and utility thresholds into arriving at management decisions, but it also involves an adaptive-management approach to doing it (see Martin et al. 2009).

State-and-transition (S&T) models (discussed in chapter 3) can be used as a basis for adaptive management provided they can be made more predictive. One way to do this is to implement them in a form that allows quantitative updating of knowledge. Bayes nets are one such example. Bayes nets are graphical models of the relationships (or causal links) between a series of variables, and the strength of the links between the variables is expressed in terms of conditional probabilities. Bayes nets are commonly proposed as tools to develop and structure process models, as they provide a method that is easily interpreted and intuitive for users, can be parameterized using a combination of data and expert knowledge, and are able to explicitly incorporate uncertainty. As an example of how you might approach doing this, the technique of quantifying an S&T model using a Bayes net was used in a trial by Rumpff and colleagues (2010) for adaptive management of southeastern Australian forests.

Adaptive Governance

Closely related to management are the rules that prescribe it. They range from behavioral and decision rules used by individuals through to the regulations imposed by various levels of government. Different disciplines tend to use the various terms somewhat differently, but from our perspective, governance is a combination of

- Institutions (the formal and informal rules including constitutions, laws, regulations, policies, behavioral rules, and norms) that mediate interactions among people, and between people and their environments

- Organizations, social networks, and the social and political processes (negotiation, suasion, information, incentives, coercion, and penalties) through which the rules are implemented

Governance (and adaptive governance) is critically important to resilience practice, but unless you're a board director or chief executive or work in the government, it's not a topic you're likely to think about. It's something that happens in the background, a tapestry of rules, rights, and regulations that we take for granted. But governance involves citizens, and private and public organizations, as well as governments, and it operates at multiple scales. Given this complexity, and because new problems, opportunities, and priorities are emerging all the time, it is clear that governance also needs to be adaptive if it is to achieve adaptive management for resilience.

Indeed, "nonadaptive" governance of a dynamic system with changing thresholds is bound to fail. Top-down, rigid, command-and-control governments are examples of being nonadaptive. Dictatorships are often like this. They might work for a while in a given situation and can even be both effective and efficient. However, nonadaptive approaches are unresponsive to changes over time or across scales and inevitably run into trouble.

Governance is adaptive when it changes in anticipation of or in response to new circumstances, problems, or opportunities. Folke and colleagues (2005) highlight the following four interacting aspects of adaptive governance of complex social-ecological systems:

- Build knowledge and understanding of resource and ecosystem dynamics to be able to respond to environmental feedback.
- Feed ecological knowledge into adaptive-management practices to create conditions for learning.
- Support flexible institutions and multilevel governance systems that allow for adaptive management.
- Deal with external perturbations, uncertainty, and surprise.

What we are calling adaptive governance also encompasses the notions of "distributive governance," that is, passing decision making down to the level in the system where it is most effectively dealt with— and this level may well change as circumstances change.

It also includes elements of "polycentric governance." Elinor Ostrom, one of its main architects, describes polycentric systems as organizations of small-, medium-, and large-scale democratic units that allow each unit to exercise considerable independence to make and enforce rules within a circumscribed scope of authority for a specified geographical area. Some units may be general-purpose governments, whereas others may be highly specialized.

Self-organized resource governance systems within a polycentric system may be organized as special districts, nongovernmental organizations, or parts of local governments. These are nested in several levels of general-purpose governments that provide civil equity, as well as criminal courts. The smallest units can be viewed as parallel adaptive systems that are nested within ever-larger units that are themselves parallel adaptive systems.

The strength of polycentric governance systems in coping with complex, dynamic biophysical systems is that each of the subunits has considerable autonomy to experiment with diverse rules for using a particular type of resource system, since each of these subunits has different response capabilities to external shocks.

In experimenting with rule combinations within the smaller-scale units of a polycentric system, citizens and officials have access to local knowledge, obtain rapid feedback from their own policy changes, and can learn from the experience of other parallel units. Instead of being a major detriment to system performance, redundancy builds in considerable capabilities.

If there is only one governance unit for a very large geographic area, the failure of that unit to respond adequately to external threats may mean a very large disaster for the entire system. If there are multiple governance units, organized at different levels for the same geographic region, the failure of one or more of these units to respond to external threats may lead to small-scale disasters that may be compensated by the successful reaction of other units in the system.

And it's not just about minimizing "bad" outcomes. Polycentric systems involving many ways of governing can lead to the emergence of successful, robust rules and "good" institutions that can spread through the system. A single, dominant system of governance doesn't learn.

Adaptive governance embraces experimentation in laws, rules, regulations, policies, plans, and investments. Unfortunately, most government bureaucracies don't embrace such approaches in their day-to-day business. Indeed, they regard them more with shock and horror.

As with adaptive management, initiating adaptive governance can be challenging, but it's an important part of putting resilience into practice. Before rejecting an adaptive governance approach, consider how your existing governance is preventing or constraining the kinds of interventions that the resilience assessment is calling for. The sorts of questions you should be asking as you contemplate adaptive governance might include these:

- How well are current institutions matched to the time scales and the biophysical, social, and economic scales at which they are required to operate?
- To what extent is adaptation of governance at regional and local scales helped or hindered by governance at state, national, and international scales?
- How can new, adaptive institutions be incorporated into current institutional arrangements?
- Can institutions be designed to be robust across a range of circumstances? Should there be "rules for changing the rules" so that institutions can be activated or silenced according to circumstances?

As a further guide, Ostrom (2009) has devised a framework for understanding the complex dynamics of social-ecological systems and identifies the following as the critical aspects of governance:

> Government organizations
> Nongovernment organizations
> Network structure
> Property-rights systems
> Operational rules
> Collective-choice rules
> Constitutional rules
> Monitoring and sanctioning processes

Any one of these can be a stumbling block to achieving resilience, and all need to be considered, along with the points already mentioned, in assessing where and how to intervene in order to achieve resilience goals.

Having read this description of the important ingredients of adaptive governance, possibly you might also be wondering if it wouldn't be easier simply to add a second garbage run every week and avoid the topic altogether. (This was actually a comment by a participant in a discussion on adaptive governance!) The language and concepts surrounding adaptive governance are dry and dense, with few direct connections and feedbacks to our day-to-day decision making, yet it serves as the template that directs all our decision making. And it's an area of resilience practice that is still opening up.

Key Points for Resilience Practice

- Appropriate actions/policies depend on the phase of the adaptive cycle the focal scale is in (as well as the phases that higher and lower scales are in).
- Consider all kinds of possible interventions—management, financial, governance, and education—not just the easiest or most obvious options.
- Consider how to best sequence the interventions you select.
- Ask yourself if your system is in a trap. If so, is transformation needed?
- How can interventions be implemented in an adaptive-management framework?
- How can adaptive governance be introduced?

CASE STUDY 4

People and Pen Shells, Marine Parks and Rules:

Why Governance Is Central to the Resilience of Coastal Fisheries

There are somewhere between 50 and 150 million people in the world who make their living through small-scale fisheries operating in coastal waters. They catch fish and harvest other marine resources, using all kinds of innovative methods. Unfortunately, despite their ingenuity, many of these operations are suffering from "the tragedy of the commons." The tragedy is the overuse of a common resource leading to its collapse.* The question is, How does a fishery agree to take no more than its fair share? What is a fair share, anyway?

According to the UN Food and Agriculture Organization, half of the world's marine fisheries are fully exploited, and a third are overexploited or depleted. The outlook for coral reefs is even worse. The *Reefs at Risk Revisited* report released by the World Resources Institute found that overfishing, coastal development, and pollution threaten more than 60 percent of coral reefs today (Burke et al. 2011). That grows to 75 percent when global pressures of climate change are included. According to the new analysis, if left unchecked, more than 90 percent of reefs will be threatened by 2030.

---

*The "commons" in the original discussion on the "tragedy of the commons," as portrayed in the famous paper by Garret Hardin (Hardin 1968), referred to the old English "common" that was shared by local herders. In fact, there were strong controls over who had access to these pastures (Dietz et al. 2003), and so the resource was not degraded as much as a true commons where open access is possible. Most deep sea fisheries and some coastal ones are true commons.

Image 7

Seri fishers with a haul of pen fish. (Photo: A. Mellado.)

It's a depressing outlook, but there are examples of some communities that are able to engage in collective action to avoid overexploitation (see images 7 and 8). One of these is the Seri fishery in the Gulf of California, Mexico.

Community Rules Make for a Resilient Fishery

Pen shells, or *callos de hacha* as the locals call them, are a very desirable catch for artisanal (local and small-scale) fishers in the Gulf of California. They are large bivalve shellfish that live buried in the sandy bottom of the gulf. Their meat is much sought after, and pen shells are one of the few marine resources with year-round demand and reliably high market prices.

Fishers dive to dig up pen shells, using a rudimentary underwater breathing apparatus connected by a long hose to an air compressor mounted on an outboard motor boat. Typically a fishing team consists

Image 8

A day's catch from a small-scale coastal fishery in Indonesia.
(Photo: J. Cinner.)

of one or two divers and two or three crew members who manage the compressor and handle the catch at the surface.

The Seri fishing village of Punta Chueca has avoided the overexploitation of pen shell stocks, whereas the neighboring fishing village of Kino has not. And yet there is much in common between them. They are located only thirty kilometers apart, and they share the same general ecosystem, harvest the same species, and use the same harvesting technology.

Underwater surveys in 2005 found densities of less than five individual pen shells per three hundred square meters in most of the Kino fishing grounds. By comparison, an average of sixty-four individuals per three hundred square meters was found at five prime Seri fishing grounds.

According to historic accounts, the Seri fishery has maintained a stable annual production for almost thirty years, fluctuating between seventy and one hundred metric tons each year. In contrast, since 1992 the catch in the Kino community has steadily declined. Production averaged only twenty tons per year from 1997 to 2003.

In the Seri pen shell fishery, most fishers find it profitable to target pen shells year-round. This is not the case for Kino, where fishers need to diversify their catch to include octopus, lobsters, and fish.

So, why has the Seri fishery proved sustainable, and why is it so different from its neighbor? The answer lies partly in the rules the Seri have created to govern their fishery and partly in small but important differences between their ecological context and that of the Kino fishery. The details in the following story were shared with us by Xavier Basurto from Duke University in North Carolina. He has been studying the Seri and their capacity to sustainably manage the pen shell fishery for many years.

Before becoming sedentary fishers early in the twentieth century, the Seri people were hunters and gatherers who occupied an extensive portion of the coastal Sonoran Desert and various islands of the Gulf of California, including Tiburón, Mexico's largest island. Today, the Seri are one of the smallest ethnic groups in Mexico. They are fiercely independent and one of the few groups in the country to avoid Spanish conquest. The federal government granted the Seri legal property rights to a portion of their historic coastal territory in the 1970s. The goal was to increase the Seri's chances of survival by reducing the probability of future conflicts with other local fishing settlements of different ethnic origin.

The pen shell fishery is managed by the Seri under a common-property regime where fishers have been able to find incentives for conservation. The entire fishery is located within the Infiernillo Channel, a narrow body of water 41 kilometers long and 5.5 meters deep on average that lies between Tiburón Island and the coast of Mexico. The configuration of the Infiernillo Channel means the Seri are able to monitor and control entrance and exit of non-Seri fishers.

The Seri have designed a number of rules with which to grant access and withdrawal rights to outsiders, who then become "authorized users." In contrast, the Kino fishing grounds operate under an open-access regime and the decision makers who make the rules are fish buyers rather than the fishers themselves. (Basurto observes that the differences in rules originally came from an effort by the Seri to be self-determining and keep out their Mexican neighbors, rather than from a strategy to avoid overexploitation of their fishing resources.)

Seri fishers determine who is eligible to enter the Infiernillo Channel as authorized fishers and which areas of the channel are off-limits to these authorized entrants. In addition, the Seri have in place a variety of monitoring and enforcement mechanisms to ensure compliance with the rules. For instance, one rule says that for a non-Seri fishing crew to become authorized users, a member of the Seri community must be hired as part of that crew. This rule confers economic benefits to different members of the Seri community because it is customary to share the catch among all the members of the fishing crew. This rule also allows the Seri to monitor—at a low cost—compliance to a rule that dictates that fishers must not fish in culturally important areas.

Culturally important areas consist of sandbars that are exposed at low tide, thereby allowing harvesting without an underwater breathing apparatus. These sites are part of a subsistence practice that is hundreds of years old. It's noteworthy because it allows women, children, and elders to participate in the harvest. The most important sandbar-harvesting events occur during the lowest spring tides and can become large communal gatherings.

To successfully harvest bivalves in sandbars during spring tides, members of the community must rely on detailed knowledge about when, where, and which sandbars are going to be exposed so that harvesting can take place during a small window of opportunity. This knowledge, and their constant presence in the channel, means that community members notice differences in abundance from one harvesting event to the next or observe the presence of unauthorized fishers.

If they notice significant or unexpected differences in abundance, they usually think (justifiably or not) that commercial fishers have been harvesting there against communal agreement. Seri commercial fishers, in turn, frequently blame outside fishers for the rule violations. An uproar then follows within the community about rule breaking, permits to outside fishers are forfeited, and the overall fishing effort in the channel decreases.

So, despite their close proximity, the Seri and the Kino fisheries differ in that the Seri have tighter control over their area and closely monitor pen shell stocks and fishing efforts. The Seri fishing area also contains extensive sea grass beds, which are largely absent from the Kino fishing grounds. The Seri avoid harvesting the shellfish in the sea grass beds because it takes more effort to dig them up and there's an increased likelihood of treading on stingrays and crabs.

This ensures that a portion of the fishing stock remains off-limits to the Seri fishery at different times of the year. The sea grass meadows are nonfishing areas in the channel that play a positive role in the regeneration of fishing stock, thus increasing the overall carrying capacity of the channel as compared with other pen shell fishing areas. The meadows are in effect providing the fishery with a buffer (a "reserve"), thereby conferring resilience to environmental disturbances combined with fishing in the open areas.

And the pen shell biology is working for the fishery as well. Pen shells are rapid growers and reach sexual maturity at one year of age. More than 70 percent of the pen shells harvested in the Infiernillo Channel are more than one year old, indicating that most of them have already spawned at least once before being harvested (Basurto 2008).

So the Seri fishers have several ecological aspects (including geography) working for them, but their institutional arrangements are an equally important part of the fishery's ongoing sustainability. Recent modeling of the Seri fishery to explore the role institutional arrangements and ecological factors play (Basurto and Coleman 2010) found that both were significant in the fishery's ongoing success. The fisher community controls who has access to their territory and actively monitors the impact of fishing activities in the channel. The pen shell resource has sufficient buffer and reserves, partly because of no-fish sea grass meadows, to enable the system to absorb a shock (like overfishing because of rule breaking) and give the human institutions enough time to reduce the fishing pressure before stocks are irreparably harmed.

Strong local knowledge, ongoing monitoring with a good institutional capacity to respond if a problem is perceived, and sufficient reserves to enable the system to respond before damage results all confer high levels of resilience on the Seri fishery.

And the management of the Seri fishery is in turn enhancing the resilience of the wider region. With its extensive sea grass meadows and mangrove estuaries, the Infiernillo Channel is probably the most important coastal lagoon in the northern region of the Gulf of California. In the twenty-five years of Seri control there has been no bottom trawling and no heavy fishing of other commercially important species. It's believed the channel actively contributes to the reestablishment and restoration of some important commercial species that have been overharvested in the past half century in the gulf region.

People and Healthy Marine Parks

The sea grass beds in the Seri fishery provide a natural form of a "no-take zone." Mostly, however, creating buffers like this requires establishment of some form of marine protected area. In many areas of the world it's the involvement of locals that is proving to be a key ingredient in the success of marine parks, such as those that protect coral reefs and coastal fish stocks.

A survey was recently undertaken of fifty-six marine reserves from nineteen countries in Asia, the Indian Ocean, and the Caribbean to see what factors might be contributing to the success of the reserves (Pollnac et al. 2010). The analysis found that three-quarters of the reserves seemed to be working, with more fish occurring inside the reserves than outside. The most successful reserves showed really big differences—with the amount of fish inside up to fourteen times the amount of fish outside; but that wasn't true in all cases. In some situations the differences were quite small, and in a quarter of the reserves there was no significant difference.

What made some reserves more successful than others? The researchers looked at a variety of factors and found social factors to be more important than biophysical aspects of the reserves. The size of the reserve and its age, for example, provided no guide to how effective it was. However, one of the best predictors of the "success" of a marine reserve turned out to be the size of the human community around the reserve, though the nature of the effect varied in different regions.

In the Indian Ocean, where reserves are government controlled and moderate in size (around six square kilometers on average), having lots of people nearby had a positive effect. That is, the reserves appeared to be working and contained higher numbers of fish inside. It is believed this could be because marine resources outside the reserves are heavily degraded, accentuating the healthier state of those inside the reserves.

In the Caribbean, however, the opposite was true. Large human populations near reserves led to poor performance of the reserves. The surveyors believed that this might be due to low compliance or poor enforcement in marine parks near population centers.

A key ingredient for a successful marine reserve was the level of poaching in the reserve, and this wasn't just about the level of enforcement. Compliance with rules governing the reserves was also related to a range of social, political, and economic factors that enabled people to cooperate better in protecting their marine resources.

Reserves worked best where there was a formal consultation process about reserve rules, where local people were able to participate in monitoring the reserve, and when ongoing training for community members was provided so that they could better understand the science (the value of the reserve to the fishery) and policy.

It was concluded that park agencies needed to foster conditions that enable people to work together to protect their local environment, voluntarily, rather than to focus purely on regulations and patrols.

Enabling people to work together to protect common environmental resources is likely the key in our struggle to govern the commons. Dietz, Ostrom, and Stern (2003) outlined several conditions that enable effective governance:

- The common resource can be effectively monitored at a low cost (appropriate feedback).
- Rates of change in the resource or its harvest aren't too great.
- Communities have good social networks enabling trust and effective communication (social capital).
- Outsiders can be excluded at relatively low cost.
- Users support effective monitoring and rule enforcement.

Few settings in the world meet all of these conditions, though the Seri pen fishery comes close. The underlying message is that effective natural resource management can only occur where linkages with the social domain are acknowledged and governance is appreciated as a cornerstone of whatever approach is applied.

Loco **Transformation**

The significance of these linkages, and the way in which they work, is nicely illustrated by the transformation in governance of the *loco* fishery in Chile, a story that is documented by Gelcich and colleagues (2010). This fishery went from a declining and degraded resource to a productive and resilient enterprise.

Economically, the gastropod *loco* (*Concholepas concholepas*) is the most important shellfish in Chile. However, in the 1980s the fishery went into steep decline. This was a little more than a decade after a military coup implemented a neoliberal fishing policy that moved the fishery from a largely domestic to an export industry, with a big increase in the number of fishers. It collapsed and was closed from 1989 to 1992 because there were virtually no *loco*. The fishery reopened during a window of opportunity when democracy returned.

Gelcich and colleagues believe that the three interlinked ingredients that allowed this to take place were (1) acceptance of the collapsed state and recognition that fishing could not continue as before, (2) an emerging scientific understanding of the ecology and resilience of the species involved, and (3) demonstration-scale trials building on experiments that identified new management pathways.

The key starting point for the transformation to the governance system that eventually emerged was the increased understanding of the role of fishers in structuring marine ecosystems. It came initially from two small experimental no-take coastal reserves administered by universities. They showed that humans controlled the abundance of *loco* populations, which in turn determined species composition in the intertidal communities.

In the absence of *loco*, the system shifts to a mussel-dominated state that has little economic value. The reserves also showed that when harvesting was experimentally restricted, resources on the seafloor, such as *loco*, sea urchin, keyhole limpet, and algae, could be restored via natural "seeding" over three to five years. This understanding created the opportunity for scientists and fisher associations to exchange information and develop larger-scale experiments, encompassing specific fishing coves, or *caletas* (Castilla et al. 1998). The experimental areas led to a learning process about stock recovery times and ecosystem dynamics, which helped to develop a shared vision of local fisher associations having exclusive rights and responsibilities to collectively manage seafloor resources.

At this same time, artisanal fishers in Chile were in the process of reorganizing into a single national confederation. It was in effect a "shadow network," an informal social network that had been suppressed by the dictatorial regime for sixteen years. It became a significant player because of its grassroots support. This support allowed the successful navigation in governance to the 1991 Fishery and Aquaculture Law, which transformed the right to fish within and between the industrial and artisanal fishing sectors. It allocated exclusive territorial users rights for fisheries, with individual transferable quotas for fully exploited species.

The key message from this retrospective analysis of a successful transformation, from a resilient but very undesirable social-ecological system to a desirable state with increasing resilience, is that there was no single solution. Legislation on its own would not have worked. The transformation required a change in legislation, a change in understanding of the ecology of the system, and a change in local-scale organization and governance. The changes influenced each other as they co-occurred. They also serve as a good example of the importance of getting past denial, creating options for change, and having the capacity for change—the three requirements for transformation discussed in the last two chapters.

Resilience Practice and Coastal Fisheries

Enforcement of fishing rules is important to sustaining coastal fisheries but this can be challenging at the best of times. It's often impossible for poor and developing countries. The resilience of coastal fisheries is enhanced where

- Fishing communities have a vested interest in monitoring and protecting their fish resource, and their governance system enables them to respond to changes in fish stocks in a timely fashion
- There is an adequate understanding of the ecology of the system and the limits to its resilience (without which it is likely that fisheries will be degraded despite legislation)
- There is an adequate buffer (in the form of a no-take zone or a marine reserve) to cushion the impacts of shocks or management mistakes

Good governance is a cornerstone of a resilient fishery.

5

Practicing Resilience in Different Ways

Depending on who you talk to, *resilience* can mean a number of things. As discussed in the introductory chapter, the four main origins of the concept lie in the fields of engineering, ecology/biology, psychology, and defense/security. Organizational resilience is now also a growing field, and it draws on the ideas developed by the other four. Resilience in economics is another area of growing interest. The literature on resilience in all these fields is large and growing.

Engineering concepts of resilience focus on a designed amount of resilience (or robustness, as they tend to call it), while in ecological, psychosocial, organizational, and defense arenas, what is important is how resilience can change—how it can be gained or lost. For engineers, it is critically important to be able to estimate the range of conditions the engineered system will have to cope with, and the design is then based on some estimate of surviving a particular threat. For example, a bridge or a dam or a nuclear reactor is designed to withstand a thousand-year flood or an earthquake of a certain magnitude. Of course, in 2011 Japan's Fukushima nuclear reactors did not withstand an unprecedented double hit of a massive earthquake followed by a damaging tsunami.

Ecologists and psychologists, on the other hand, embrace uncertainty and assume that there will always be surprises; that real uncertainty doesn't allow for the prescription of the range of conditions the system must cope with. Resilience is about coping with both known disturbances *and* unknown and unexpected disturbances. Embracing uncertainty involves recognizing that it keeps us on our toes, stops us from getting stuck on narrow pathways, and acts as a positive influence on the way we live and plan for the future.

With its rising popularity and wide use in mission statements, *resilience* is being increasingly applied in many different arenas. In this chapter we consider how it might be applied in different kinds of problems, and how the insights from one area might inform applications in others. The discussion is structured under six problématiques, three where resilience has traditional roots and three (health, the law, and economics) where it doesn't but nevertheless plays a role.

This is not a comprehensive overview of different uses of resilience ideas, but it does highlight some of the diversity of its meaning and application.

Problématique 1: Psychosocial Resilience, Identity, and Coherence

Why is it that in a group of people subjected to the same traumatic experience, some cope well while others suffer terribly?

Psychology is one of the oldest roots of ideas on resilience. In this arena *resilience* is commonly defined as the positive capacity of people to cope with stress and catastrophe. One definition is about the ability to bounce back to homeostasis after a disruption (return to normal). But it is also about how adaptive systems use exposure to stress as a way of enhancing the ability of people to cope with future negative events. The connections to concepts of social-ecological resilience are easy to see.

The mainstream psychological view of resilience is defined in terms of a person's capacity to avoid psychopathology despite difficult circumstances, and the central process involved in building resilience is the training and development of adaptive coping skills.

Resilient people and communities are more inclined to see problems as opportunities for growth. Resilient individuals seem not only to cope well with unusual strains and stressors but actually to experience such challenges as learning and development opportunities.

While some individuals may seem to be more resilient than others, it is important to recognize that resilience is a dynamic quality, not a permanent capacity. Resilient individuals demonstrate dynamic self-renewal, whereas less resilient individuals find themselves worn down and negatively impacted by the stressors of life. Some have an inherently higher resilience capacity than others, but in all people resilience is dynamic. It can be enhanced and lost.

Resilient people are not necessarily "good" people. We may heroize people for their capacity to bounce back, but that doesn't make them better or morally superior. Resilient people can be found in all walks of life—the good, the bad, and the ugly. The concept of psychological resilience has no direct relationship with morality. As with intelligence, it all depends on how it's expressed.

In recent decades, building psychological resilience in at-risk populations has become an increasingly important target of community intervention, youth work, and personal development programs. For example, resilience is a key theme in the forty developmental assets for children and adolescents developed by the Search Institute (http://www.search-institute.org/). Community efforts to enhance resilience through intervention have been increasingly proactive, preventative, and potentially cost saving. Enhancing psychological resilience seems to be an underlying theme in challenge-based personal development programs such as Outward Bound (http://wilderdom.com/obmain.html).

Psychology researcher Ann Masten has studied resilience in children growing up under conditions of disadvantage and adversity. Her most surprising conclusion is the ordinariness of resilience (Masten 2001). An examination of the findings suggests that resilience is common and that it usually arises from the normative functions of human adaptational systems, with the greatest threats to human development being those that compromise these protective systems. The conclusion that resilience is made of ordinary rather than extraordinary processes offers a more positive outlook on human development and adaptation.

A Sense of Coherence

A valuable insight into psychosocial resilience comes from the work of Astier Almedom on refugees in displacement camps in Eritrea (Almedom et al. 2007). The camps were set up to deal with internally displaced people (IDP) resulting from the war with Ethiopia from 1998 to 2000. The war, between two of the world's poorest countries, killed tens of thousands of people and displaced many times more.

Almedom used a "sense of coherence" index in order to measure an individual's resilience. The index was based on the work of medical sociologist Aaron Antonovsky. He explained human adaptation and response to extreme psychosocial stressors in terms of a mobilization of "generalized resistance resources." These resources include genetic predisposition, acquired skills, and accessible social support. His analyses led him to believe that people managed stress and stayed well because of the strength of their sense of coherence. A sense of coherence is the characteristic of understanding the world around us, feeling up to challenges confronting us, and believing the challenges are worth taking on.

Almedom used her adapted sense-of-coherence scale to assess resilience in 265 people, living in two groups: nondisplaced urban residents and people living in IDP camps. Participants were asked a structured set of questions so that resilience could be assessed in quantitative terms.

The nondisplaced residents from villages showed higher scores for coherence than displaced people in the camps. However, if a displaced person came from a mobile community, such as a group of nomadic pastoralists, then the scores weren't that different from those for the nondisplaced residents. In other words, mobile pastoralists in the displacement camps coped well (were resilient). Their identity was not linked to a particular place or village. Those who were not coping tended to be people who had lived all their lives in one village and lost their identity when moved into a displacement camp.

Almedom's work in Eritrea suggests that attachment to place (and deriving identity from this) is critical for farming communities and not so important for mobile pastoralists. Identity, she believes, is also about a way of life that is tied to modes of livelihood. Strictly speaking, the pastoralist way of life has a broader link with place—the seasonal migrations from lowland to midaltitude involve adaptation and identifying with more than one place. Those with mobile modes of livelihood are naturally more adaptable, with more than one place to call home. It's a complex picture with intricately interwoven adaptive social systems.

People go through a psychosocial transition following a crisis (such as displacement), and whether the outcome is positive or negative depends on the kinds and levels of emotional, cognitive, and material support they receive. Almedom makes the point that this support needs to be both micro (local, family) and macro (government, NGO)—which again emphasizes the role of cross-scale connections in resilience.

Themes of sustaining a desirable identity, flexible institutions, adapting to change, and understanding the context of the system resonate with many of the elements of a resilience practice.

Problématique 2: Disaster Relief and Crisis Management

How do resilience concepts relate to the capacity of communities to cope with disasters?

The security/defense arena is another where resilience has a tradition. B. W. and a social science colleague, Frances Westley, were involved in a meeting of the Ditchley Foundation (in the United Kingdom) in 2009, on "Society's Resilience in Withstanding Disaster." It brought together academics, politicians, bureaucrats, and people involved in international disaster programs to discuss the role of society's resilience in helping cope with disaster (Walker and Westley 2011). Here we present some insights from those discussions.

Speed of Return Does Matter: Fast, but Not Too Fast

Ecological resilience is more about the capacity of a system to recover following a disturbance than the speed of that recovery. It's about crossing, or mostly trying not to cross, thresholds, rather than quickly returning to the equilibrium. Many ecologists consider the *speed* of recovery as being much less important. Indeed, it's seen more as a distraction from the critical issue of whether or not the system *can* recover. However, at the Ditchley meeting there was general agreement that after a major disaster, the longer a community stays in a disturbed state, the more difficult it becomes for that community to recover. Eventually it may not be able to recover.

Being in a disturbed state sets up secondary effects that erode the system's capacity to self-organize and respond. So, length of time in a disturbed state may itself be a variable with a threshold. Many participants placed much importance on this, as they had witnessed what became of a community that was left too long in a state of disrepair—consider the case of New Orleans following Hurricane Katrina or Port-au-Prince following the Haiti earthquake.

A rapid response is therefore important, but a quick fix usually isn't the answer. In some situations a short-term response (e.g., a three-day recovery effort) is not sufficient to "trigger" recovery, as the problem is too complex to engage with in that time frame. In such situations the "short-term, quick fix" response might in fact be causing a kind of peripheral blindness, a preference to focus on only those disasters that are conducive to a quick response, and a tendency to declare victory (as President Bush did in the Iraq war) on a superficial basis.

Meanwhile the longer-term vulnerabilities that result from disasters are ignored, passed over to other services and agencies, moved elsewhere, and never measured when adequacy of a disaster response is considered.

Perceiving Problems

Based on involvement in counterterrorism activities, one participant proposed the "Murphy's water bed" principle, to wit, if you push down a problem in one place, it pops up somewhere else where you don't expect it. It is therefore sometimes better to leave a potential problem where you can keep an eye on it, and learn about it. But it is still necessary to be able to respond to unexpected issues in unexpected places, which relates to the problem of trading off general resilience against specified resilience. Though it is sometimes best to keep things specific, we should be organizing resources to expect the unexpected and not neglect general resilience.

One interesting observation was that the severity of disasters, and hence the risk associated with different kinds of disasters, is measured in terms of individual "body count," or mortality. As the total risk measured this way is 100 percent (as "no one gets out of here alive"), it becomes a highly political process of claiming which percentage is associated with which risk. The key question, it was thought, should not be, What is the proportional risk associated with which threats? It should instead be, Which disasters are more likely to reduce overall resilience of the system?

Response Origination

Where should responses to disasters best originate? Government representatives expressed the need to push power up, to the international level, in an attempt to anticipate and provide adequate response to threats such as terrorism, which seemed to have a truly global dynamic, *but* at the same time to push power down to the local community level, where sense making, self-organization, and leadership in the face of disaster were more likely to occur if local governments felt accountable for their own responses.

One discussion framed the need in terms of promoting the philosophies of both Hobbes (a social contract, ceding freedoms to a higher authority) and Rousseau (looking for the good in people to develop personal responsibility from the bottom up). This is reflected in the social-ecological resilience literature on the need at local scales for adaptive governance and comanagement, and at higher scales for global-scale institutions in the face of looming global-scale failures (something discussed in chapter 6). Current efforts and emphases are focused too much at the levels in between.

Westley (personal communication) has recently done some further analysis in this area of disaster resilience and has come to the following conclusions:

- Crises are often viewed, by those interested in transformation, as windows of opportunity.
- In crises/disasters, dealing with particular threshold effects may depend on hierarchy, speed and accountability, and heightened attentiveness.
- However, the capacity to respond creatively to specific crises depends on general resilience, and general resilience depends on consultation, engagement, learning, sense making, empowerment, and social justice at all scales.
- To build resilience, we need to understand its sources at individual, organizational, and community levels.
- A culture of learning, not blaming, needs to accompany approaches to complex system change.

There are interesting synergies between work on anticipating disasters and work in social-ecological resilience. The disaster work sheds light on the difficult role of government agencies in disaster mediation, the problematic role of accountability, the importance of time—not too long (time as a threshold) and not too soon (quick-fix failures)—in achieving an adequate response, and the importance of sense-making capacities at the community level.

One way resilience thinking can inform disaster studies is through its capacity to see normal times and times of disaster and collapse as different phases of the same system. Another way is through the appreciation of controlling (slow) variables as an important and overlooked part of most risk assessment. Resilience theorists, on the other hand, can also learn from the many examples experienced by the disaster relief community.

Problématique 3: Engineering

Do ideas on resilience from ecological, psychosocial, and disaster arenas have anything to contribute to engineering?

As we said earlier, engineers take a different slant on resilience. They prefer the term *robustness* and, as is appropriate for engineers, it has a design connotation.

Here we want to introduce some new ideas in engineering that are converging on resilience thinking from other areas—a kind of "meta-robustness" approach. To do this, it is informative to consider how the role of response diversity (or designed redundancy, as it is in this case) has been used in engineering. We are indebted to John Doyle (California Institute of Technology and the Santa Fe Institute) for the following example.

Aircraft designers place a lot of emphasis on ensuring that aircraft keep on flying when things go wrong, or when they are hit by unexpected shocks. As you'd expect, a lot of effort goes into the mechanical design of the aircraft to make sure that not only is it aerodynamically efficient and built with ultra-high-quality components but that it will also keep on flying and be controllable if bits stop working or get knocked out.

However, the real complexity lies in the control of the aircraft. This involves sophisticated feedback systems that make the whole system far more robust than its component parts. Airplanes like the Boeing 777 have many parts, some three million of them provided by five hundred suppliers around the world.

The control systems have to be very robust. Both their hardware and software components have to keep functioning, come what may. The simplest way to ensure this would be to put some replicas of the control system in the plane, and if the one that was operating got knocked out or failed for some reason, a reserve could be turned on. But that's a very blunt way to think about redundancy, and in the case of the Boeing 777 (and other aircraft), designed robustness has been introduced in a more considered manner.

Four independent teams were engaged to develop the software that controls the thousands of parts in the control system. All of the designed systems have to do the same thing—perform identical functions—but do it in different ways and hopefully not have any common bugs.

To help ensure things are done differently, the teams aren't allowed to talk to each other and a lot of effort goes in to picking the teams so that they come from different places, have different educational backgrounds, and work with different computer languages and debugging tools. It's all done to get as much diversity as possible despite the products' having the identical function.

The 777 story illustrates that diversity is an important resource and you get most bang for your buck in robust design by combining components with different features rather than through redundancy per se. But it's not foolproof, and since we can't afford the time frames and high losses associated with evolutionary selection, some engineers are now thinking about how to implement designed robustness in a somewhat wider context, in a kind of meta-robustness approach.

So, where a conventional engineer might try to build a robust system like an airplane, a resilience engineer might in addition worry about the design environment the engineer is in, and whether there are problems with the institutions that might make it hard for the engineer to make good trade-offs. It's about the robustness of the robust design process to the assumptions, tools, infrastructure, and institutions of the designer.

All of which amounts to a convergence of this meta-robustness approach with resilience ideas. It has started to move away from the assumption that the variance the system has to confront is bounded and known and to move toward something more akin to the idea of general resilience.

Problématique 4: Resilience and Health

Being resilient is commonly seen as an attribute of good health, but how resilient is the health system itself?

Health systems worldwide seem to be in crisis. Rising costs are spreading services thinly, and there is controversy about the relative costs/benefits and equity in private versus state systems. The one point of agreement is that many are worried about the resilience of the health system as a whole in the face of a health crisis. Health science has not been one of the roots of resilience, but professionals involved in the health system are becoming increasingly involved in the use and application of resilience ideas. Here are two examples of the interplay between resilience and health.

The Obesogenic Crisis

The crisis of obesity is a mix of the "long fuse, big bang" and "ramifying cascade" surprises (see the discussion on surprises in chapter 2). For decades now, obesity has been on the increase in almost all developed countries. Its causes are many and complex but chiefly come down to an increasingly poor diet, overconsumption, and lack of exercise due to changes in patterns of mobility. Processed foods dominated by fats and carbohydrates have become cheaper than fresh, high-quality food. They are easily available in all cities and increasingly dominate our diets. This is especially so in disadvantaged groups. The advertising and food industries have compounded the effect.

The rise in levels of obesity has been a slow variable that has escaped the notice of many people. The worry is that when it reaches a critical level (which it is doing) there will be secondary feedbacks to the rest of the health system. Beyond that point there are significant consequences for the welfare and resilience of society.

A UK government program called Foresight suggests that being overweight is now the norm and that by 2050, 60 percent of men and 50 percent of women could be obese. Obesity increases the risk of a range of chronic diseases, particularly type 2 diabetes, stroke, and heart disease. It's also linked to cancer and arthritis. The UK National Health Service estimates that costs attributable to conditions connected to being overweight and obese are projected to double to £10 billion per year by 2050. The wider costs to society and business are estimated to reach £49.9 billion per year (at 2012 prices).

Such is the scale of the problem that it is frequently referred to as an obesity epidemic. Our human biology is being overwhelmed by the effects of today's "obesogenic" environment, with its abundance of energy-dense food, motorized transport, and sedentary lifestyles. As a result, the people of the United Kingdom (as one example of an advanced country suffering from the epidemic) are inexorably becoming heavier simply by living in the Britain of today.

The pattern of change is typically an increase in the body mass index (a measure of body fat based on the ratio between height and weight), with an associated increase in the risk of diabetes and other diseases. Above some critical level, the likelihood of diabetes becomes very high, and this is where the ramifying cascade comes is. Once a person is diabetic, the likelihood of contracting other diseases, notably various cancers, rises sharply. This represents an irreversible threshold because even if the people concerned subsequently lose weight, they remain diabetic. And diabetes is a very costly disease.

There are feedbacks connected with the obesogenic crisis throughout the health system. Because health is an immediate priority, the funding shortages it generates spread to other areas of public expenditure. In Australia today, for example, the health system takes up more of the budget than the education system. With rising health costs, the situation promises to deteriorate. That has serious ramifications for the future.

As people become more obese they are less inclined to exercise, and the problem gets worse. It is compounded by seemingly helpful, but actually perverse, actions in society. Some shopping malls, for example, now provide free motorized scooters for disabled and obese shoppers. They may attract obese customers, but walking around the mall would clearly have a better social outcome.

The obesity epidemic cannot be prevented by individual action alone. It demands a systems analysis and a societal approach. The Foresight project says tackling obesity requires far greater change than anything tried so far, and at multiple levels: personal, family, community, and national.

Foresight's work indicates that a bold whole-system approach is critical—from production and promotion of healthy diets to redesigning the built environment to encourage walking, together with wider cultural changes to shift societal values around food and activity. This will require a broad set of integrated policies including both population and targeted measures and must necessarily include action not only by government, both central and local, but also by industry, communities, families, and society as a whole.

In many ways it's similar to climate change. Both need whole societal change with cross-government action and long-term commitment. There are also many direct connections. Our colleague Tony Capon, from the National Centre for Epidemiology and Population Health at the Australian National University, points out the cobenefits that will arise from tackling climate change and obesity. Measures such as increases in cycling and walking, reducing traffic congestion, and shifting to more of a vegetarian diet would have positive effects in both areas. He extends the idea to the redesign of cities to facilitate advances in both.

The Human System, Resilience, and the Law

Tim Buchman is the founding director at Emory University's Center for Critical Care in Atlanta, Georgia. Like all trauma units, Tim gets people who arrive in his surgery close to death. They have sustained massive trauma—in accidents or through intentional violence—and unless treated quickly they will die. In complex systems jargon, they are already in the death basin of attraction but haven't yet reached the attractor (which is death). Tim's job is to stop them from reaching that state and try to get them back into the "alive" basin of attraction.

As he explained, his unit swings quickly into action, aiming to stabilize the vital systems of the body, pumping blood into it if needed and then maintaining the correct blood pressure, and regulating the body temperature, the blood pH, the amount of oxygen the person is administered, and several other things that are prescribed by the protocols of several expert professional societies.

Once this is achieved, the trauma team pauses to observe what happens. Sometimes the patient begins to recover. Other times, she or he doesn't.

Tim observed that during the waiting times, while stabilized, patients' bodies were always trying to change; the blood pressure would drop a little, or the breathing rate would increase a bit, or the body temperature would change. The traditional protocols demanded that these changes be corrected immediately. But Tim wondered, What if these changes were reflecting feedback processes that had evolved to maintain body functions? Mammalian bodies are, after all, self-regulating systems (as evidenced by our ability to maintain a constant temperature under a wide range of external temperatures). Tim relaxed the previously stringent guidelines, allowing his patients far more latitude in adjusting their own physiology.

Around that time, an interesting report emerged from Houston. In a study of injured patients, two immediate care protocols were compared. One, the traditional approach, required immediate resuscitation to hemodynamic end points (basically, blood pressure levels). The other minimized resuscitation until after the source of bleeding was identified and controlled. The result was a significant increase in survival in the group allowed to self-regulate.

There are two lessons from this experiment about resilience, and they relate to two scales. The first is a confirmation that a human body is indeed a self-organizing system. Allowing a body to probe its resilience boundaries enables it to self-organize and maintain or increase its resilience.

The second lesson is that cross-scale effects can influence resilience in a different kind of way. Until this study was reported, no surgeon or hospital would be prepared to take the risk—even though the patient would have had a higher probability of living.

The need for good protocols is, of course, clear, as is the need for them to be subject to the law. This is where the law serves to ensure standards and legitimacy. But tort laws are now such that not only do they inhibit adaptive practice by professionals like surgeons, but people, generally, are encouraged to look for legal compensation for all manner of things where in fact they should have looked out for themselves. And the role of some law companies whose business falls into what is commonly referred to as ambulance chasing has elevated risk averseness to levels that undermine resilience of the health system; and the health system needs to have a level of flexibility and adaptiveness.

The tendency to protect and to avoid exposure to risk can have all sorts of secondary consequences that lower resilience. For example, there are several published trials showing that children exposed to dirt (bacteria and other microbes), like children reared on farms, are less likely to develop asthma than those in cities or those not allowed to play in dirt.

In the kinds of cases described above, an unintended consequence of the law is that it is acting to reduce resilience in and of society. The question for the law fraternity is, How should the law, as a sector of society, respond to this problem?

Problématique 5: Resilience and the Law

What is the relationship between the law and resilience? And how does resilience thinking apply to the practice of law?

Interaction between resilience and the law is a mixed bag. In some areas, where lawyers are working with ecologists and other scientists to improve outcomes, interaction can build resilience. In others, as illustrated in the example of the human body, and in the example we describe below, it can have unintended negative effects.

Many of the conflicts that arise between the law and a resilience perspective stem from a fundamental difference in their premises. A resilience approach advocates the need to probe the boundaries of resilience in order to maintain it and recognizes the need for flexibility and adaptiveness. The law, on the other hand, is risk averse, fearing the consequences of creating precedents. This results in one-size-fits-all solutions and little capacity for adaptive responses.

A resilience assessment is full of contingency and needs to be very adaptive; the law needs to be tight and explicit to avoid interpretations that subvert its intent. The problem is bedeviled by issues of scale, reflected in the relationships between the different courts that operate across the full spectrum of the law. As explained by our colleague Andrew Edgar (University of Sydney), if a specialist court is established by legislation (such as the Land and Environment Court in New South Wales in Australia), a generalist court will always have a supervisory jurisdiction over it (in NSW, it is the NSW Court of Appeal). The supervisory jurisdiction is limited but sufficient to make sure the specialist court does not do anything too radical or progressive (such as heavily favoring public and environmental interests over the private interests of a developer!).

Land courts, appeal courts, supreme courts, and all other courts have their own purposes and see a particular issue differently. Administrative courts are those that deal with administrative law, the law that governs the activities of government agencies. Since administrative law is intended to achieve legitimacy, it should have a positive effect on resilience. However, it doesn't always work out that way.

Sometimes the secondary consequences of something that is clearly of immediate and direct benefit can have wider effects that negatively affect others. Consider the U.S. Endangered Species Act. It leads to intense conflicts between conservationists and some other sectors, and when one considers that in developing policy for the Rio Grande the focus of practically all action is on two declared species, it's possible to see why other sectors might object.

It would appear that the law and resilience are in conflict because of two trends that are readily apparent in Western countries—risk aversion, which reduces resilience, and the rising propensity to shift blame and look for legal compensation rather than accept responsibility for one's actions. Secondary effects of risk aversion come back in large ways. There is a need to probe the boundaries of resilience in order to maintain it (consider the human body example), and risk aversion is inimical to that. The blame shifting is related to this too, because it again takes away the need to learn from one's mistakes. Examples help to illustrate this, and we present one to do with water and the law—in particular, the water that serves as the lifeblood of the Everglades in the U.S. state of Florida.

Sandy Zellmer and Lance Gunderson have examined how legal structures and interventions strongly influenced the pattern of development, and therefore of resilience, in the Everglades (Zellmer and Gunderson 2009). The Everglades is a rich mosaic of waterways, saw grass prairies, mangroves, and cypress swamps. It's a story of declining biodiversity, complex cycles, and thresholds involving nutrient levels in the water.

Today, national legislation governing water quality, biological conservation, and the production of resource outputs plays a significant role in efforts to restore natural values in the Everglades. The basic playing field is dominated by concepts of maximum sustained yield (MSY)—maximizing whatever it is the system is expected to deliver. This is despite the fact that approaches underpinned by the MSY concept were never really a viable policy—because they never really worked. It was prevalent for a long time, and surprisingly, it's still in vogue in many places.

Recent restoration efforts in the Everglades are being driven by environmental protection laws that appeared back in the 1970s. In particular there is the Clean Water Act and the Endangered Species Act, as well as a landmark in U.S. conservation law—the National Park Service Organic Act.

In 2008, a federal district court opinion in an Everglades court case (one in a long series of court cases) found that the Environmental Protection Agency had ignored the requirements of the Clean Water Act for satisfying water quality standards when it approved the state's revised schedule for cleaning up phosphorus-laden water flowing into the Everglades from Lake Okeechobee. Special treatment marshes to filter the polluted runoff in water flowing south started to be built, but in 2003, the sugar industry pushed a bill through the legislature that replaced the 2006 deadline with a gradual schedule of benchmarks that don't begin until 2016. The court described the new law as "an adroit legislative effort to obscure the obvious" in creating an escape clause that allows noncompliance.

As a result of these and other delays wrought by the ongoing legal tussle, the cost of cleaning up the water is now estimated at over $10 billion, with completion not anticipated until the 2030s or 2040.

Zellmer and Gunderson's paper provides details on several other pieces of legislation, becoming ever more complex and intertwined, and in their words, "the Everglades restoration effort continues to be characterized by ever more planning rather than any on-the-ground action."

In a separate assessment of adaptive management based on the lessons from the Everglades legislative saga, Gunderson and Light (2006) suggest existing statutes governing endangered species recovery, national park management, and pollution control need regulatory reform (not dismantling) to ensure their adaptive implementation to achieve their goals. But the extant Flood Control Acts require a complete overhaul to mandate adaptive management and ecological restoration and to move the Corps of Engineers away from the unbounded cost-benefit analysis currently in place.

Whether existing requirements are supplemented, modified, or rescinded, new legislation will require more than just a mandate that adaptive management be pursued. Just stating that adaptive management should guide restoration has had virtually no success in the Everglades. Gunderson goes on to suggest that legal vehicles should enhance flexibility, learning, and adaptive approaches, rather than reinforce pathologically resilient institutions and ecosystems.

The failure of the ongoing legal saga of the Everglades throws out a challenge: How do you meet the needs for legitimate accountability without reducing resilience? Perhaps it might help if those involved in proposing legislation, before it is drafted, put a resilience lens over the proposed legislation and took a long-term, systems view of the secondary effects the proposed law would have.

The tension between the dynamic adaptive way social-ecological systems work and the need for the law to be tight and explicit will always be there. More attention to resolving the tension would benefit the law (and society). However, the rising risk averseness in society, reflected in the law, seems an avoidable trend that is having perverse outcomes.

Problématique 6: Resilience and Economics

Given that economic policies patently influence the resilience of social-ecological systems, does a resilience perspective offer any valuable insights into economics?

Our aim here is to briefly reflect on how resilience thinking intersects with economic thinking. We are indebted to our colleague Buzz Brock for many of the details in the following discussion. Brock is a prominent economist from Wisconsin and a leading light in the world of resilience science. Which leads us to an important introductory point—there is much overlap between the world of economics and resilience thinking.

Stability and robustness analysis, for example, is a major theme in economic studies. This type of analysis is closely related to resilience because it seeks to understand the response of economies to disturbances and whether economies can fall into alternate stable states with unpleasant consequences. One might say that Keynes, for example, became famous for his analysis of unpleasant alternate stable states in economics and what to do about them. Two of the three founders of modern general equilibrium theory (a central pillar of economic theory), Kenneth Arrow and Lionel McKenzie, have written extensively about stability analysis of economic dynamics (Arrow and Hahn 1971; McKenzie 2002). A recent treatise on robustness analysis in economics is Hansen and Sargent 2008.

A major feature of resilience analysis is the role of slow variables that may degrade resilience of the system to shocks. There has been much recent activity in economics that focuses on slow variables, which are hard to observe or tend to go unnoticed and which can cause unexpected abrupt changes in economic systems (e.g., Hommes and Wagener 2009).

In the following discussion we focus on two ways in which resilience has relevance in economics: the economics of resilience, and the resilience of the economic system itself (our main focus).

The Economics of Resilience

In the sense of empirical estimation, the economics of resilience is about how much the inhabitants of an economy are willing to pay for extra resilience. This field is in its infancy and there have been only a few attempts to estimate the value of resilience.

Maler and colleagues (2007) have proposed treating the amount of resilience in a system as a "stock" that can be valued in terms of its contribution to well-being. Using this approach allowed for an estimate of the value of resilience to rises in the water table and consequent salinization of land in an estimate of the "inclusive wealth" of the Goulburn-Broken catchment in Australia. In this case it was relatively straightforward, since the amount (stock) of resilience could be equated to the distance from the water table to two meters below the soil surface. If the water table reaches this two-meter threshold, the water, and the salt in it, are drawn to the surface by capillary action.

The alternate states of the system on either side of this threshold—healthy soils on one side, unproductive salinized soils on the other—have dramatically different economic values. This approach, therefore, allowed for the estimation of how much it was worth investing in maintaining the stock of resilience—that is, keeping the water table below two meters from the surface.

Not many social-ecological systems have such clearly demarcated thresholds, and valuing resilience is an area needing considerably more work. Furthermore, the economics of resilience in the sense of measured willingness to pay for enhanced resilience has only been considered at local scales. If we scale up to the world, then the question involves the costs/benefits of crossing/not crossing planetary boundaries. Though it is an important question, it will not be easy to provide unambiguous estimates.

The Resilience of the Economic System

The global financial crisis (GFC) and its aftermath have raised concerns (yet again) among economists and noneconomists about the instability of the global economic system, and its resilience is therefore an important issue.

In "The Financial Crisis and Economic Policy" (Campbell et al. 2011), Benjamin Friedman points out that the financial system cost around 10 percent of all profits earned in America thirty years ago, but it now consumes about 30 percent. He asks whether it is doing its job well and whether it is worth this much in benefits to the economy.

In the same paper, Robert Solow makes the argument that as the financial system increases in size and complexity, it contains a large potential for instability at the same time its lobbying power is proving very influential. Given the lobbying power such a wealthy industry can muster, Solow points out, it can be effective in preventing useful regulation of this industry. Politics and lobbying clearly play a significant role in this story. Many commentators were worrying about the increasing fragility (i.e., increasing lack of resilience) of the financial system before the GFC. But, as McCarty and colleagues (2010) document, politics and ideology neutered many attempts to usefully regulate the system.

Attributes of resilient systems, as discussed in this book, have direct application to economics at fine scales. Portfolio diversity is a basic principle of investment, for example. It's much the same as response diversity in an ecosystem or social system. At the scale of macroeconomics and growth theory, it's more complex. We need to start by considering the purpose of the economy.

The economic system is a social construct, developed to serve society. Solow, one of the originators of modern economic growth theory, defined a sustainable economy as "non-declining per capita human well-being over time." That says nothing about the need for economic growth per se, and yet the focus of attention in economics has become centered on growth—dominated by the absolute need for the economy of a country to grow in order to avoid rising unemployment and its dire consequences. Countering that mainstream economists' view, there are arguments made by many (including economists) for an uncoupling of growth from resource consumption, or even the need for "de-growth."

Others make the point that we actually need to stop worrying about growth and worry about what matters to society. Jeroen van den Bergh, for example, says that we need to think about an "a-growth" economy, rather than growth or de-growth, because the focus needs to be on how to get the economy to achieve its purpose (van den Bergh 2011). Its purpose is long-term per capita human well-being. Satisfying peoples' aspirations for being employed (being needed, and so forth) is part of that, but on the other hand, if the economy causes indirect changes in society or the environment that lower human well-being, then it's not doing its job properly. And there are many examples of that.

One case is known as the rebound effect or the Jevons paradox. Jevons was a nineteenth-century economist in England who noted that efforts to increase the efficiency with which coal was used (deemed necessary because it was thought England would run out of coal) actually led to more coal being used, not less. It became cheaper to use for all sorts of purposes, and so it got used more and more.

Scaling this up, if unintended secondary effects lead to the world exceeding planetary boundaries (as described in the next chapter), then the economy most certainly will have failed in doing its job, because human well-being will then significantly decline.

Another example of what one might call "economic growth malfunction" is the problem of positional goods, stressed by Robert Frank in his book *Luxury Fever* (Frank 1999). Frank argues that much of economic growth ends up in a positional-goods race where, for example, a homeowner loses satisfaction with his house when houses around him become bigger. Yet another example of economic malfunction is the growth of income inequality and the problems of "the winner takes all" where fewer and fewer people end up with more and more wealth, relative to the rest of the population. This leads to a rather unpleasant society to live in.

Top economists who study the matter seriously debate whether growth in incomes beyond a certain point increases measured happiness and measured life satisfaction in advanced economies (Easterlin 2003; Kahneman and Deaton 2010). While economic growth and increased incomes tend to enhance life satisfaction in poorer societies, the above findings by scientists who work in this field show rather ambiguous results for incomes beyond a certain point and for growth beyond a certain point.

Australian economist John Quiggan makes the point in his book *Zombie Economics: How Dead Ideas Still Walk Amongst Us* that the economic system keeps repeating the same mistakes (Quiggan 2010). Fallacious theories are proven wrong during crises, buried, but then rise again. It seems they can't be killed off. The economic system does not learn from its mistakes and crises. That's the kind of pathological resilience that keeps a system in an undesirable regime. Healthy resilient systems are learning systems that allow adaptation and change as circumstances change.

Quiggan raises four main objections to the current economic system, based on the repeated resurrection of four failed tenets of economic theory:

- The efficient markets hypothesis (that markets are the best guide to investment and production): Pushing this led to (demanded) financial deregulation, removal of controls of international capital flows, and so forth, which enabled the conditions leading to the GFC.
- The validity of using computable general equilibrium models to run economies: These models are able to integrate the interactions of all markets to explore the consequences of policy decisions. But they have a series of problematic assumptions built into them, such as implied efficient markets, the use of representative agents, and rational behavior in consumption and production, and they don't deal with discontinuous changes in resource supply (threshold effects).
- Trickle-down economics: The "rising tide lifts all boats" sounds good, but the trickle-down effect has repeatedly been shown not to work. It actually leads to the rich getting richer and the poor getting poorer.
- Privatization: The repeated failure of governments and the private sector to deal with the problems of market failure in privatized enterprises has tempered enthusiasm for privatization all over the developed world. Comprehensive privatization failed so badly that it now does seem to have really died. The challenge is in developing the right mix of public and private ownership in a mixed economy.

In his final chapter, "Economics for the Twenty-First Century," Quiggan addresses concerns about the resilience of the economic system by emphasizing the need for a new approach to risk and uncertainty. Given the recent collapse of yet another economic "New Era," he says, economics should focus (1) more on realism, less on rigor; (2) more on equity, less on efficiency; and (3) more on humility, less on hubris. (We note in the final chapter that many people believe the quality of humility is important for a resilient world.)

Because arguments about free markets are so common, and frequently heated, it's worth expanding a little on this. Markets will deliver the best of outcomes provided some critical conditions are met: First, everything that is important is included in the market (e.g., there are no unpriced ecosystem services). Second, there are no property rights problems (e.g., who owns the fish in the open oceans?). Third, everyone involved and affected is fully informed and has equal access. And finally, the market is fully cleared. Seldom if ever are these conditions met, and the job of governments is to address the consequent market failures. How well they do this will strongly influence the resilience of the economy.

A common criticism of economics is that it is too reliant on theory and elegant mathematics (Quiggan's point 1). However, it's all too easy to be critical of economics and economists. We shouldn't throw out the baby with the bathwater. The world needs markets and a largely free and bottom-up system because they spawn experiments and novelty, and technological progress, which we do need. Central planning doesn't work. Consider the communist states before the fall of the Berlin Wall, and Mao's China. A mixed economy is what is needed. The change from a centrally planned economy to a market economy (and vice versa) represents a transformational change in an economic system. They are fundamentally different kinds of economic systems.

Within the market economy system, the challenge is getting the right balance, especially how and when the government needs to intervene to correct market failure, and a resilience approach would suggest it cannot be a fixed balance; the balance needs to be adaptive.

Beyond national scales, the world picture suggests that global corporations and nations are driven by self-interest and competition, with little allegiance to global-scale institutions that will ensure common good at the global scale. Climate change is the obvious example, but there are other issues that are equally worrying, such as increasing antibiotic resistance and nuclear proliferation. And the big problem we face in the resilience of the global economy has everything to do with human aspirations.

As noneconomists, we'd like to avoid stamping around where angels fear treading. The science of economics has long recognized the problem of alternate stable states and has devoted a lot of effort to the study of mechanisms that promote good stable economic states and avoid bad stable states, as well as mechanisms that enhance or impede the responsiveness of the economy to shocks and disturbances. But as we have illustrated throughout this book, resilience places an emphasis on interactions and threshold effects in the whole human-natural system, and analyzing one subsystem (like the economic system) has to take into account the likelihood, and consequences, of significant (irreversible) shifts in the others—the natural, technical, and social systems with which the economic system interacts.

Those who *study* economics do that, but it's not so common among the many who *practice* it. The same can be said, of course, about the theory and practice of ecology—the gap between the two is reflected in the many examples of degraded ecosystems, just as in economics it is reflected in the many examples of unwanted societal and ecological outcomes, including GFCs.

CASE STUDY 5

Out of the Swamp:

Lessons from Big Wetlands

The swampy wetlands of the world have played important roles in human history. The presence of water made them favored places for animals and people alike, and they have long histories of human use and manipulation. And it's the manipulation that is of interest from a resilience perspective, since it has invariably resulted in unintended secondary effects. We begin our swamp tour with a discussion of two major wetland systems in the developing world: the Okavango Delta in Botswana and Tonle Sap in Cambodia (see images 9 and 10). Both are very valuable assets for the countries involved, and both depend on the continued flow of water from countries upstream.

Facing Change in the Okavango Delta

The Okavango Delta is an inland delta in southern Africa. It is one of the largest, least-disturbed delta ecosystems in the world, covering an area of over seven hundred thousand square kilometers when flooded. It has been listed as a Ramsar Wetland of International Importance and is globally significant for its biodiversity. And the wildlife brings in the tourists, making the region an important source of revenue for the surrounding people. In Botswana, tourism is the second largest earner of foreign currency, and this is largely based on the Okavango Delta.

The system is fed by the Okavango River, which has a catchment

Image 9

The Okavango Delta, Botswana, where the wildlife comes to the water and the tourists come for the wildlife. (Photo: S. Walker.)

in nearby Angola. The river flows across the Caprivi Strip in Namibia and enters Botswana in the northwestern corner of the country. Downstream from here it is confined in a narrow depression known as the Panhandle, but from the town of Seronga it spills out over a large area as it divides into a number of distributor channels, forming a vast alluvial fan called the Okavango Delta.

Each January a giant pulse of water from heavy summer rains over the south of Angola enters the Okavango River and begins a five-month journey through Namibia to the richly biodiverse swamp in Botswana's semiarid savanna. The flooded area expands and contracts in response to that pulse.

The core of the delta always has water cover; this is the "permanent swamp." But the swamp expands to three times that size when the water arrives between June and August. At the fringe there are the "seasonal floodplains," and some areas, known as "occasional floodplains," are flooded only during very wet years. These three types of

Image 10

Tonle Sap, which sustains the economy and ecology of Cambodia and neighboring Mekong countries. (Photo: L. Ruettinger.)

wetland—permanent, seasonal, and occasional—differ in biogeochemical processes, vegetation, and animal populations, thus providing a range of different services for people.

Around six hundred thousand people live in the basin. They rely on its waters for small-scale agriculture and livestock, fishing, and household use. But aside from evaporation, a few sips drawn off to supply the Namibian town of Rundu, and 1,100 hectares of irrigation nearby, the majority of the water that falls in Angola at the turn of the year arrives in Botswana in midwinter (June) to recharge the delta.

And that makes the Okavango River a rarity in that it has hardly been touched by human development along its 1,100-kilometer length. Shaping its future is the delicate task of the Okavango River Basin Commission (OKACOM). The intent is that any country that wants to develop its part of the basin must go through consultation and investigate the impacts of that development on the river flow or the ecosystem.

Under Pressure

But, as with every important wetland system in the world, there is continuous and growing pressure on the river and its delta. When Namibia faced severe drought in the late 1990s, it considered drawing water off the Okavango to supply its capital, Windhoek, hundreds of kilometers away. Namibia also has a long-standing desire to build a hydroelectric dam on the river at Popa Falls, fifty kilometers upstream of the border with Botswana.

Further north, the consolidation of peace in Angola means a growing population around the river's headwaters, and the government in Luanda—flush with money from oil—is turning its attention to long-delayed rural development.

And, within Botswana, there is rising pressure on the edges of the delta as the human population increases. In this case the benefits from tourism (based on biodiversity and a healthy basin ecosystem) need to flow to local people, and this is occurring to some degree.

Overarching all these country-specific challenges is the specter of climate change, with the expectation of increased variability in rainfall and greater water scarcity.

As might be expected, Botswana opposes any additional extraction of water from the Okavango River, arguing that it will disturb the ecology of the delta, leading to lost biodiversity and revenue from tourism. But what can Botswana offer Angola and Namibia to secure water?

In the face of mounting socioeconomic pressures, Angola, Namibia, and Botswana signed the OKACOM Agreement in 1994. Its aim is to develop an integrated management plan to ensure the sustainable future of the basin and its associated ecosystems. It commits the three countries to manage the Okavango River basin so as to promote coordinated and environmentally sustainable regional water resources development, while addressing the legitimate social and economic needs of each country.

What Are the Important Questions?

From a resilience perspective, what are the important aspects of this system? It's best to approach this question at a number of scales. It doesn't matter from which direction you start, but let's start at the community/ecosystem scale. At this scale we begin by asking if there might be any ecological threshold effects that require priority attention. One obvious example concerns the response of the Okavango ecosystems to changes in river flows and associated water regimes and alterations in the quality and quantity of sediment loads entering the delta. The major change of concern in controlling variables of the delta is the amount of water that flows into it. In advance of any proposals to divert water or forecasts of reduced water through climate change, a prime resilience question is, As water inflow declines, are there any critical threshold levels that, if passed, would lead to significant and/or irreversible changes in ecosystems?

What threshold effects might occur, at different scales, in different parts of the delta? The answer to this question would provide essential information for development strategies, both within Botswana and at international levels. Answering the question requires a good understanding (preferably a good integrated model) of water dynamics and the response of vegetation and animals to changing water regimes, at a number of scales.

For example, the permanent swamp in the delta has eight vegetation communities, determined by channel dynamics, flow variations, hippo effects, and fire (Ellery et al. 2003). What are their limits to coping with water regime changes before they can no longer recover? Not all of the eight communities will be able to persist under different water regime scenarios; some will change into others, and some will disappear as certain threshold levels are reached. Is there a threshold water regime below which the core swamp essentially disappears—changes into some other kind of ecosystem, with too little marsh to support viable populations of the essential marsh species?

For comparison, a recent assessment of the river-red-gum/box woodlands of Australia's Murray River system has shown that, owing to water extraction for agriculture all along the river and combined with recent climate changes (and forecast climate change), most of these woodlands have passed a threshold and are now dying. These river-red-gum ecosystems cannot recover. The management priority is now how to make the remaining ones, the forests that are still viable, more resilient (NSW NRC 2009).

Coming back to the Okavango, in addition to water-related thresholds, there are questions that need to be posed concerning critical threshold levels in regard to areas of, and connections among, required habitat types for different functional groups of animals.

Other issues are revealed by asking questions about other kinds of shocks the system is, or may be, subjected to. There are some exotic plant species—potential pests—in the system, both aquatic and terrestrial weed species. Are there threshold environmental levels (water or soil related) associated with the likelihood of any of these crossing into pest outbreak modes?

Because the Okavango's biodiversity supports a booming tourism industry, there are opportunities for community-based natural resource management. Community trusts run businesses based on the natural resources (thatching grass, wildlife) of the local environment. At this scale, are there minimum size/number requirements for conservancy areas (tourism license) to support a viable tourism operation?

In 2006 there were thirteen community trusts in the Okavango region and over forty villages were involved, some supported by local or international NGOs. This policy of involvement of local people in the tourism industry enhances the general resilience of the system, as a social-ecological tourism-based system.

The Okavango Delta system is at a critical point in its history. The coalescence of regional peace and economic development is placing competing demands on the water that is essential for its survival. Without a multiscale, multinational resilience assessment of its linked biophysical and social dynamics, including the relative costs and benefits of the unavoidable trade-offs that are involved, it is likely that the problems will be dealt with in a partial manner. And this will result in the unintended secondary consequences that inevitably follow implementation of partial solutions.

The Tonle Sap: Sustaining the Heartbeat of a Nation

The Tonle Sap is a combined lake and river system lying in the central plains of Cambodia. When full, the Tonle Sap is the largest freshwater lake in Southeast Asia and one of the most productive fisheries on the planet.

Like the Okavango Delta, Tonle Sap's natural character is shaped by a massive annual flood. What distinguishes Tonle Sap from most other wetland systems is that this flood pulse is a two-way event.

For most of the year the Tonle Sap Lake is fairly small, with an area of around 2,500 square kilometers and depth of about one meter. During the monsoon season, however, the lake more than quadruples to cover up to 15,000 square kilometers with a depth of up to nine meters, flooding surrounding fields and forests.

The lake usually empties into the Mekong River via the 120-kilometer-long Tonle Sap River. But during the southwest monsoon the water level in the Mekong rises so fast that part of the floodwaters run back up the Tonle Sap River, causing the river to reverse its flow back toward the Tonle Sap Lake. Thus the lake's only outlet for a time becomes just one more inlet. Once the flooding has passed, the river flow reverses again and the lake again empties into the Mekong River. The reversal of the Tonle Sap's flow acts as a safety valve that prevents flooding farther downstream in the Mekong. Later, during the dry season (December to April), the Tonle Sap Lake provides around half of the flow to the Mekong Delta in Vietnam.

This exceptional water regime provides an enormously important range of ecosystem services to the people of the region. In addition to the flood mitigation and water supply, the seasonal inundation of the floodplains around the lake makes the system one of the most productive freshwater ecosystems in the world.

The flooded forests and fields on the floodplains offer excellent shelter and breeding grounds for fish, and there is extensive migration of different fish species between the Tonle Sap Lake and the Mekong. During the inflow there is mostly a passive migration of eggs, fry, and fish to the Tonle Sap Lake and its floodplains. Later, large numbers of fish follow the receding floodwater back to the lake and finally back to the Mekong River. The lower fish migration route, between Tonle Sap and Vietnam's Mekong Delta, provides food security and the livelihoods of many millions of people.

The lake, therefore, is a unique aquatic habitat of great biological diversity, supporting important genetic resources for the region and an immensely productive inland fishing resource. Fisheries from the Tonle Sap provide supplies of fish not only to Cambodia but also to neighboring Mekong countries. This is of great nutritional, economic, social, and cultural importance. Fish provide up to three-quarters of the animal protein in the average Cambodian's diet. They generate employment and income.

Upstream of the Heart

Because it beats in and out with the rise and fall of the Mekong floods, revitalizing the whole region as it does so, Tonle Sap is often referred to as the "heart" of the Mekong, an organ critical to this region's life cycle. And just as the heart is inextricably linked to the body that it nourishes, so too Tonle Sap is linked to and shaped by what's happening in and around it.

The list of challenges facing the system is long and growing. There are multiple developments in upstream countries, such as dams and irrigation schemes posing threats to water supply, with enormous potential to shift the hydrological regime that has formed this unique social-ecological system. On top of this there are the potential impacts of climate change.

The socioeconomic setting of the people on and around Tonle Sap is complex and challenging. The area is experiencing rapid population growth, there is widespread poverty, and most of the people in the area are deeply dependent on the lake. On top of this there is considerable ethnic diversity and unequal access to natural resources.

Keskinen and colleagues (2007) examined several case studies of the challenges being faced by different groups of people dependent on Tonle Sap as things change. One involved the ongoing loss of flooded forests due to firewood cutting and conversion of the flooded forests into agricultural land. The flooded forests form a key element of Tonle Sap's ecosystem.

On top of this there is development planned in the upstream countries—most notably the construction of large hydroelectric dams in China and Laos. These are likely to cause an increase in the dry-season water level in the lower parts of the Mekong, and consequently in the Tonle Sap Lake. The rise of dry-season water level means an extension of the permanent lake area and thus changes in the floodplain. The most notable change would be permanent submersion, and subsequent death, of remarkable areas of remaining flooded forests.

The reduction of flooded forest area would mean the loss of livelihoods for a significant number of people, due to both the loss of flooded forests and the consequent negative effects on aquatic production. Thus, increased development in other Mekong countries would have negative effects for the ecosystem and livelihoods of the Tonle Sap floodplain, potentially fueling additional conflict in the area.

Dealing with transboundary impacts on Tonle Sap is difficult at the best of times, but it's being made even more challenging because of China's growing economic dominance in the wider region and the dependence of Cambodia on China's economic cooperation and assistance.

Changing Rights on Floodplains and Fisheries

Another emerging challenge is the growing privatization of the lake edge. Traditionally, large parts of the floodplains—particularly those close to the lake—have not been under clear ownership or cultivation. They have been considered common property.

The drive for agricultural production along with increased land value has led private investors—often belonging to the country's emerging elite with connections to investors elsewhere in Southeast Asia—to see the floodplain areas as profitable targets for investment. The increased flow of investments to Tonle Sap's floodplains has resulted in a rapid expansion of irrigated agriculture and related structures such as large embankments and reservoirs primarily for profitable dry-season rice cultivation. A significant proportion of new irrigation areas and structures may actually be illegal.

Keskinen and colleagues (2007) describe how the emergence of private irrigation areas in the Tonle Sap floodplain has meant that many local communities have lost areas that they have traditionally used for floating rice cultivation and as grazing grounds for cattle, thus undermining local customary rights. As these areas are usually not officially titled to villagers, the villagers have found it difficult to resist the development. The conflict over the use and control of floodplain areas is thus closely linked with the broader governance context and its ambiguities.

Just as there are tensions over who has rights to use the floodplains, so too are there problems with access to fishing territory. This is especially the case with the operation of large-scale, commercial fisheries that are based on a fishing-lot system. Fishing lots are geographical concessions auctioned to the highest bidder for a certain period, usually two years. The owner of the fishing lot has an exclusive right to harvest fish from the lot, to sublease parts of the lot, and to keep everyone else out.

The system excludes most people from the most productive fishing areas during the most productive fishing season. This has created serious tensions and even armed conflict between local villagers and fishing-lot owners. Responding to the growing conflict, the government proclaimed radical changes in 2001 in which half of the total area of the private fishing lots was changed to public fishing lots open for community fisheries. Although this eased tensions, management of community fishing areas has turned out to be challenging. Institutional arrangements of community fisheries often seem to ignore the complexity of local power structures.

Clearly, the resilience of the Tonle Sap system depends strongly on the kind of governance system that is evolving as Cambodia moves into the modern world. Whether it succeeds in meeting the governance challenges it faces, and whether it can match Ostrom's (2009) eight critical requirements (see chapter 4), will determine whether its current identity can persist.

As with the Okavango system, the Tonle Sap system is at a critical point in its history. A number of ecological thresholds need to be examined and properly understood so serious, and possibly irreversible, changes can be avoided. The capacity to deal with these biophysical resilience issues depends on the social adaptability of the system, at the scale of the lake system itself, and the big issue of changes in the flow regime will be determined by higher-level cross-scale (transboundary) connections and trade-offs.

Staying Resilient in the Swamps

Both the Okavango Delta and the Tonle Sap are inherently very resilient ecological systems under the environmental regimes in which they evolved. The traditional social-ecological system that has developed in each region has also proved to be resilient to fluctuating water conditions and other shocks, based on common property rules. However, both are now threatened by novel environmental conditions imposed by upstream users and growing pressures associated with climate change. International agreements recognize this, but how strong are they?

A resilience practice approach to the problem suggests that an appropriate framework for engaging with these challenges requires stakeholders to

- Include the whole catchment as "the system," with multiple scales of concern (the countries being one)
- Take a whole-system view of the situation, identifying important feedbacks between different parts of the system and across scales
- Provide very clear information on the ecological and social costs and benefits (in both directions) of changes due to upstream water use, including information on critical levels/changes in the water regime associated with particular, significant losses (threshold effects)
- Get the institutional arrangements to match the governance needs, including involvement of other international stakeholder interest groups (NGOs, United Nations bodies like Ramsar, World Wildlife Fund, and so forth) to strengthen and refine the agreement

Lessons from Other Swamps

As a contrast to the Okavango and Tonle Sap systems, consider the Camargue wetlands (discussed in chapters 2 and 3), the Macquarie Marshes in the Central West catchment of New South Wales (discussed in case study 3), and the Everglades (discussed in chapter 5).

The Camargue is at a very different stage of development in response to human use, with different issues, and therefore calls for a different approach to putting resilience into practice.

The state-and-transition "model" of the Camargue was developed in an effort to resolve the competing demands for how the delta system should be managed. The big difference between it and the Okavango and the Tonle Sap is that the flow regime of the Camargue has been thoroughly interfered with for centuries. Long-established levees and diversion banks have been used to regulate water levels for various purposes. As figure 9 shows, through various management interventions the ecosystem can be shifted into different states that suit, or do not suit, those who compete for them—livestock breeders, duck hunters, reed thatchers, conservationists.

There are no transnational issues (the region is all within France), and so the emphasis is on understanding how management can influence the areas in each kind of state, the relative values of the ecosystem goods and services provided by each state, and how to get agreement on the trade-offs among them. This is a big challenge. Failure to get agreement is common in situations like this all around the world, and the failure often leads to deteriorating management and ecological conditions.

A resilience approach helps because it identifies critical levels where small extra gains by one side may cause very significant declines for others. The French group involved in this kind of problem has developed a procedure for engaging the different stakeholder groups, using a model that the stakeholders themselves develop (with the aid of the scientists) and then use to investigate how different management or policy options meet their wishes. In doing so, the stakeholders also see how their preferred management options influence the preferences of other stakeholders.

The modeling platform they have developed (called Cormas; Bousquet et al. 1998) allows for the easy development of an ecological model (ComMod) that captures the essential dynamics and can produce such outcomes as figure 9. This is then coupled with a role-playing game (an agent-based model). In the Camargue example this game is called BUTORSTAR (Mathevet et al. 2007). It engages the stakeholders in trying to achieve desired outcomes but requires them to negotiate with each other as the results of their decisions unfold. It is played in a series of steps, allowing discussion and negotiation as the players learn about the consequences of their decisions. It therefore allows the group to work toward a management policy that all can live with.

Methods like BUTORSTAR require time for data collection, model development, stakeholder engagement, and playing the game. The importance of arriving at a lasting, equitable solution determines whether the investment is worth it. In the case of the Camargue it seems it is. And it can become an ongoing learning process. Mathevet and his colleagues observed that the different "players" adopted different styles of negotiation and conflict resolution and grouped these into five modes of dealing with conflict (based on Thomas and Kilmann 1974). These are competing, avoiding, compromising, collaborating, and accommodating. This brings social dynamics very much into resilience practice.

In terms of development history and intensity, the Macquarie Marshes are somewhere between the Camargue and Tonle Sap. The water regime here is all-important and is in large part determined by the management of outflows from Lake Burrendong (a dammed reservoir) upstream. The dam's primary purpose is storing water for irrigation. The lesson for resilience practice is the need, once again, to consider all scales as well as cross-scale interactions. Because they have been the dominant players, and especially during periods of normal rainfall, the irrigation operators have not considered downstream consequences of how and when they release water into the river that supplies the Macquarie Marshes.

Among the issues identified in the resilience workshop on the marshes was the problem of introduced carp. This exotic fish is causing drastic reductions in native fish species. In trying to identify the main causes for this, it became clear that an important factor is the temperature of the water released from the dam. The water is drawn from the bottom of the dam and is very cold—too cold for native fish species to breed but not too cold for carp. So for the first hundred kilometers or so below the dam, carp have complete control, and by the time the river reaches the marshes, they are overwhelmingly dominant.

Taking the water from near the surface of the lake, where it is warm, was never considered. However, though there would be some costs involved in re-engineering, those attending the workshop considered that this option needed to be revisited, since it would reduce the carp's advantage and contribute to enhancing the resilience of native fish.

Managers of the Maquarie Marshes wouldn't be able to predict with any certainty what might happen to fish ecology if such re-engineering of the offtake system were undertaken. What's required is some structured ecological experimentation, some learning in an adaptive management sense.

Of course, such an approach may require some resourcing, but more important than money is the cultural attitudes of the governing bodies across each of the scales—a willingness to engage with and learn. Attempting greater control by throwing money at the problem without such a cultural shift will likely fail. Or worse, it could make the situation even more intractable. For proof of that you only have to look at the Everglades, one of the most expensive swamps in the world in terms of management and restoration.

Despite the injection of billions of dollars from the federal government, the Everglades seem to simply lurch from one crisis to the next. Efforts at restoration continue, yet everything seems to be held in gridlock by litigation (see chapter 5). It's an example of a pathologically resilient management system (Gunderson and Light

2006), not only unable to effectively deal with the changes being experienced but not able to change the management in response to this failure. It's trapped.

Since 1990 a major focus of management in the Everglades has been toward ecosystem restoration. In 2000 the Everglades Restoration Act was passed authorizing almost $8 billion for restoration purposes, and adaptive management was explicitly put forward as a means of meeting restoration goals. Since then, however, conventional planning approaches have been the main approach, with no implementation of structured ecological experimentation, central to adaptive management. Gunderson and Light (2006) describe the culture of management in the Everglades as more focused on resolving past conflicts than discovering sustainable futures. Management continues to focus on planning and seeking certainty (a certainty that is illusory) prior to action. Gunderson and Light argue that the Everglades needs to seek a transition to adaptive governance, as a way of increasing responsiveness, and to generate more diverse and versatile competencies that create options for the future.

In warning about what could happen to an important swamp if we fail to adapt and learn from crisis and opportunity, it's difficult not to include the specter of the devastated Aral Sea (a tragedy described in chapter 4). Its rulers applied top-down, command-and-control management, ignored key slow variables, focused on flows while ignoring stocks, and always believed they could dig themselves out of an ever-deepening hole by using technological fixes. The outcome was that the Aral Sea was transformed from a prosperous and resilient social-ecological system to an economic and ecological basket case.

If we think of these six iconic water regions (swamps/lakes/deltas) as lying on a continuum from best condition, and ecologically most healthy and resilient, to most degraded, then the Okavango Delta lies at the best-condition end and the Aral Sea at the most degraded. All had their own attributes of biophysical resilience that enabled them to persist for thousands of years in their original states, changing and self-organizing in response to climatic fluctuations and initial human use. Some in the middle range are now quite different in proportional composition and perhaps inherent productivity to what they were in their original state, but most still have the same basic identity as functioning social-ecological systems. The Aral Sea does not, and it was governance, based on a short-term perspective that failed to recognize this social-ecological system as a self-organizing system with dominant feedback effects, that led to its current parlous state. And it is governance that will decide the trajectories of the other five regions described in this case study.

Lessons from Swamps for Resilience Practice

Four common messages emerge from analysis of the changes occurring in the five swamps discussed in this case study and in the Aral Sea, discussed in boxes 3 and 5:

- Attempting to manage the system by focusing on only the scale of the wetland is likely to fail because wetlands by their very nature are strongly connected to what's happening in the surrounding region and upstream. A whole-system perspective is essential.
- It is necessary to provide clear information on the ecological and social costs and benefits (in both directions) of changes due to upstream water use.
- Governance is crucial in maintaining the resilience of a healthy wetland. Guiding governance through periods of political and regional change is a challenging task.
- Adaptive management is essential for exploring the complexity of wetlands.

Whether or not the Okavango Delta follows an Aral Sea trajectory will depend on the extent to which resilience principles are incorporated into water management policies and practices across the whole river basin; and that will depend on the effectiveness of governance.

6

A Resilient World

W hat does *resilience* mean when applied at the planetary scale? Our planet as a global system has changed quite a bit over the last ten thousand years, but when viewed on geological time scales it has been very stable during this period, a time known as the Holocene. Indeed, despite some changes it's had the same identity.

As a species we've developed in an environmental space defined by climate and natural ecosystems that have proved remarkably resilient. We've cleared native vegetation, reduced the variety of life, changed the balance of nutrients in rivers and ocean basins, and significantly altered the very composition of the atmosphere. As all of this has gone on, the global system has responded in a self-regulating way—by storing more carbon in the oceans, for example—that has enabled it to continue to function in the same way. The big feedbacks that regulate our planet—in particular, energy flows, water cycles, food chains, and nutrient cycles—have continued to behave in much the same way that they have over the last ten thousand years.

It appears, however, we are now close to a situation where this could change, that the global system within which we have developed is close to crossing planetary boundaries that could lead to the world moving out of its stable Holocene regime.

In this chapter we reflect on those boundaries, the type of governance needed to engage with this challenge, and a view of climate change from a resilience perspective. We then return to a discussion that we initiated in *Resilience Thinking*—what are the attributes of a resilient world?

Planetary Boundaries

A group of scientists from many disciplines recently identified nine "planetary boundaries" that they believe we can't afford to ignore (Rockström et al. 2009). Staying within these boundaries keeps us within the Holocene environmental regime—the environment that has enabled human societies to develop from primitive to what they are today.

The boundaries relate to critical transition points in the states of the world's climate, atmospheric chemistry, and marine and terrestrial chemistry (nutrient states and cycles, and acidity) and rates of biodiversity loss.

The study concluded that safe operating limits have been exceeded for three of these nine boundaries—the rate of biodiversity loss, nitrogen inputs to the biosphere and oceans, and climate change. The group also believes that limits are being approached for two other boundaries (stratospheric ozone depletion and ocean acidification), that there is a need to take urgent action on three others (phosphorus cycles, change in land use, and freshwater use), and that insufficient information exists to assess the other two (atmospheric aerosol loading and chemical pollution).

The actual levels of these boundaries have lots of uncertainty about them, and several are not thresholds in the strict sense we use in this book. We are well aware of the limitations in proposing those boundaries (B. W. is one of the paper's authors) and also aware of the criticisms that have been leveled at the paper, though some of these are due to misconceptions. For example, in regard to biodiversity, the boundary that has been crossed is the *rate* of loss of biodiversity, not the actual loss so far. The world cannot sustain the current rate of loss of species and genotypes without eventual drastic changes in ecosystem function.

This can be illustrated by an example where an actual loss of species is causing a drastic change in ecosystem function. Fishing pressure has already resulted in significant loss of species on many coral reefs, and when these reefs are subjected to inflows of nutrients, they flip into algal turf systems and do not recover even if nutrients decline. In the absence of large herbivorous fishes, the algal turf prevents the reestablishment of coral polyps. Furthermore, the threshold amount of nutrients that causes this flip gets lower as water temperature rises, so the resilience of coral reefs to fishing and nutrient shocks will decline as global warming occurs.

For most of the world, biodiversity loss is not yet critical as it is on coral reefs. But if the current rate of biodiversity loss continues, then more and more of the world's ecosystems will experience analogous changes. If we can drastically reduce the rate of biodiversity loss, we can return to the "safe" side of the biodiversity loss boundary. But that is not what's happening. By 2010, the International Year of Biodiversity, all the countries that had signed the Convention on Biological Diversity were supposed to demonstrate a significant reduction in the rate of biodiversity loss as measured in 2002. But rather than decreasing, the third Global Biodiversity Outlook noted in 2010, extinction rates were increasing everywhere.

Some of the planetary boundaries are real thresholds, like ocean acidity. Below a critical pH level, many forms of plankton cannot develop their tiny calcified shells, because the concentration of aragonite (the form of calcium carbonate they need) is too low. Most higher marine organisms (and our fisheries) depend on a healthy plankton layer in the food chain.

The planetary boundaries concept will inspire a lot of research and analysis and will be refined over time. For example, the original assessment did not take into account freshwater accumulation and flows; a re-analysis of the phosphorus boundary, which does do this (Carpenter and Bennett 2011), shows that the phosphorus boundary has been exceeded. It highlights the spatial differences in freshwater eutrophication arising from differences in use of phosphorus fertilizer and suggests that recycling phosphorus from regions of excess use to regions of deficiency could mitigate eutrophication and help reverse the boundary transgression.

Collectively, the nine boundaries mark safe limits for the planet as we know it. Decision makers should appreciate and acknowledge that

- Long-term human well-being is very likely to decline if they are exceeded
- They are not independent—crossing one can increase the likelihood of crossing others
- The closer we get to each boundary, the smaller the shock it takes to push us across

A wrong interpretation of the boundaries idea is to assume that everything will be fine right up to each boundary and, so as long as we don't cross them, we can continue pushing up to the edge. The fear that people will interpret it this way has been one of the criticisms of the paper by some scientists. However, as stressed in the third point, the prudent thing is to stay as far away as possible from the boundaries.

In the absence of the human drivers forcing the world toward the boundaries, it is likely that Earth would continue as it is for many more millennia. However, if we cause the Earth to transgress the boundaries, it's a different story. The planetary system will certainly continue to self-organize, but it will do so along a different trajectory. Life on Earth will certainly exist long into the future, but there's no guarantee that it will include us humans.

We do not intend to discuss all of the boundaries here. The Rockström et al. paper does that, though we suggest you read the full-length paper published in *Ecology and Society* (2009) and not just the summary *Nature* paper. That three of the boundaries have already been crossed and others are very close is a clear indication that the "Anthropocene," the age we are now in, is unlikely to continue on its current trajectory. Whether the boundaries are reversible is not really known, and discovering their true nature should be a priority task.

In thinking about resilience at the global scale, it is once again necessary to highlight the importance of general resilience. The identification of the planetary boundaries has generated a lot of interest, controversy, and discussion about the world's safe operating space, all of which is good. But most of this discussion comes from the perspective of *specified* resilience, how resilient particular aspects of our global system are to specific threats.

How do we consider *general* resilience at the global scale? One way, perhaps, is to use an analogy from the other end of the scale, the scale of our genes. Molecular geneticists and evolutionary biologists talk about the concept of "evolvability"—the ability of a population of organisms to generate genetic diversity and then evolve through natural selection. It involves the potential for genes to change (mutate) into different forms and so provide new opportunities for natural selection and new adaptations. It also involves having unused (redundant) bits of genetic material that can be brought into play if an opportunity arises. It makes the process of evolution robust.

How much evolvability does the world have? How much unplanned novelty are we creating, and losing? How much redundancy and unused material do we have at the global scale to bring into play when conditions demand that we come up with something new?

As unexpected disturbances hit us, how responsive will the world be? What new combinations/uses of the world's resources will be available to us? Evolvability writ large should get our focused attention, along with the specified planetary boundaries.

Global-Scale Governance

How is it that the world is right up against so many planetary boundaries (and may have crossed some)? Is it simply ignorance? Probably not; some of our best minds have been grappling with these issues for decades.

Is it because of willful, short-term greed? That's part of the problem for sure, but it's not all down to greed because, again, considerable effort has gone into designing systems and structures of governance that take vested short-term interest into account.

Of course, the reasons behind how we got into this situation are complex. But one important consideration, recognized by many, is the tension between the sovereign imperatives of nations, which drive competition between them, and recognition that nations need to cooperate. It harks back to an important insight from resilience studies that building resilience at one scale can reduce it at other scales, most usually the scale above. And this is the tension we need to examine, the tension between resilience at national scales and at the global scale.

However, before we do that, it is important to recognize that a loss of resilience can also occur in the other direction (from higher to lower scales). Global-level developments, in particular the suite of things that go under the banner of "globalization," can lead to the loss of resilience at local, national, and regional scales. Here, it is the global-scale aspirations of globally connected multinational corporations, attempting to secure their own resilience, that erode the resilience of social-ecological systems at finer scales, worldwide.

A message that has become increasingly strong as we've worked through the process of how to put resilience into practice is the role of feedbacks in determining threshold changes. Managing for resilience is about managing critical feedbacks. It has also become clear that many of the critical feedbacks are those that operate across scales. Cross-scale effects are often the most important and often the least recognized in linked multiscale social-ecological systems. Possibly the feedbacks that are most threatening the resilience of humanity at a global scale are (1) from national- to global-scale dynamics, (2) from the global scale to subglobal dynamics, and (3) among the global-scale changes themselves.

The enormous advances made in society and in human well-being since the Industrial Revolution owe much to the operation and the power of nation-states. Since the Hobbesian days when life was "solitary, poor, nasty, brutish and short," allegiance to a sovereign power has enabled people to get on with advancing their lives.

The combination of the agricultural/industrial revolutions and the rise of nation-states has worked so well that there are now over seven billion people on Earth and we have entered this new era, the Anthropocene, in which human activities are strongly influencing the functioning of the planetary system. And we now face some looming global changes that are very worrying.

The world is now truly acting as an interconnected system in almost all ways. The global financial crisis, and its recurring hiccups, demonstrate just how interconnected the economic system is—the modularity that conferred resilience has been eroded in favor of increased financial performance of global financial markets.

The biggest challenge we face is that the actions of nation-states have effects on the functioning of the global system, and these effects, which feed back to effects on nations, are not recognized, are ignored, or are simply not being dealt with. Competition between nation-states overrides and precludes the cooperation that is necessary at the global scale.

An analysis of global governance in *Science* (Walker et al. 2009) found that the global institutions that do exist tend to address the concerns they were set up to deal with largely in a silo fashion. For example, the World Trade Organization deals with trade, the World Health Organization with health, the Food and Agriculture Organization with food and agriculture, the United Nations Environment Programme with the world's environment. In other words, though they refer to each other in the ways in which they operate, they mostly ignore the connections to the other silos and the feedback effects that occur among them.

The nongovernment commercial institutions that operate at the global scale—the global corporations—are driven by quarterly shareholder profits and so also ignore the secondary feedback effects on global functioning, because for the most part their consequences are far enough away to be ignored in the next annual report.

Yet it is the unrecognized feedbacks between all these institutions that are leading to the world's closing in on the planetary boundaries we do not want to cross.

While there have been recent encouraging signs that the sustainability of coastal fisheries under the control of individual nations is improving (e.g., Ostrom 2008), all the open-ocean fisheries are in rapid decline, with many approaching critical thresholds (as occurred in the Grand Banks cod fishery). And this is despite international agreements, which unfortunately have been largely unenforced.

Adaptive governance is now gaining ground at subnational scales in many developed-world countries, which is good since it is a necessary component of general resilience. Top-down, one-size-fits-all governance doesn't work, and neither does pure bottom-up activity. An adaptive system of distributed governance is necessary for the social-biophysical feedbacks to be managed in a timely and effective way. At the global scale, it is glaringly absent. Our largely ineffectual attempts to manage the feedbacks involved in climate change are the most obvious example.

The question is, Can we continue to adapt and finesse our way forward, or is a regime shift to an alternate global state of the world now inevitable unless we undertake transformational changes in governance?

Global governance needs priority attention by national as well as global leaders. We're not advocating a global government. It's probably impossible to achieve, and it would be too risky—we can all think of national leaders who we definitely would not like to see heading it. What's needed is an integrated system of global governance that takes the interests of the globe as priority while addressing the trade-offs between nations and the globe. The United Nations does many good and necessary things, but it does not fill the governance need, since it is not about governance. Its mandate is about missions of various kinds, but mostly security, authorized by the UN General Assembly or Security Council. In any case, it isn't one organization that is needed; it's an interacting set of agreements and institutions that collectively provide the necessary governance.

Can we undertake a graceful transformation to avoid planetary-scale consequences before transformation is done to us in a thoroughly unpleasant and ungraceful way? As identified in the set of determinants of transformability, the first attribute is acknowledging the need for a transformational change. We are currently, predominantly, still in the state of denial. The world as a whole does not yet accept that we need a system of adaptive global governance to which nations and global corporations pay allegiance. Yet we patently do.

Climate Change and Resilience

Climate change is the clearest example of national interests su-
perseding, and undermining, global resilience. It is not within the
scope of this book to undertake a detailed account of the issue of
climate change, but we will make three points that arise from a
resilience perspective. They arise from a starting point based on
an acceptance of the scientific consensus that global warming is
occurring and is due to increasing atmospheric concentrations of
greenhouse gases.

First, there are threshold levels of greenhouse gas (GHG) concentrations (and therefore changes in climate patterns) that will cause regime shifts of many kinds, at many scales, all over the world. Some will
be more significant than others, and a few will cause large, cascading
effects on others. The 350 parts per million "safe" level of carbon dioxide that has been advocated is just that—a safe level that will prevent
any major, serious regime shifts.

Second, given that resilience is the amount of disturbance a system can absorb without crossing a threshold, there are two ways in
which resilience can be lost, or gained. One way is that the state
of the system can change, in terms of the slow, controlling variable. That is, the amount of GHG can increase, bringing the state
of the system closer to the threshold. As the levels of GHG rise,
resilience is declining. Conversely, if GHG declines, resilience will
increase. What this means is that as GHG increases, it will take
progressively smaller climate shocks (weather events/patterns) to
push the many kinds of systems across the many kinds of thresholds around the world.

The other way resilience can be lost or gained is by changes in the
positions of the thresholds—changes in the levels at which GHG causes
feedbacks to change, and therefore a system regime shift. For example,
decreases in overall species diversity and in landscape connectivity
will result in a system flipping at a lower level of GHG.

The practical implications of these two ways of losing resilience suggest two actions: (1) do what you can to promote reduction in GHG,
and (2) understand what determines the positions of the thresholds
that are of concern, and therefore how to move them so as to increase
the resilience of what is of concern.

The third point that arises from a resilience perspective is that GHG level (and therefore climate change) is one of the planetary boundaries, and it will interact with the others (à la the figure 6 matrix) in ways we have not yet been able to work out. Managing resilience of the global system needs this information.

Attributes of a Resilient World

What would a resilient world look like? It was a question we posed at the end of *Resilience Thinking*. We proposed nine attributes of a resilient world. They weren't principles as such, more aspirational goals. In this book we've revisited these ideas and attempted to describe ways in which these attributes might be developed. These attributes are listed below. Some relate to specified resilience (such as acknowledging slow variables), but most of them are more connected to the general coping capacity of a system and are therefore aspects of general resilience.

1. Diversity: A resilient world would promote and sustain diversity in all forms (biological, landscape, social, and economic). Can you identify trends where declining diversity might lead to a change in some important feedback? Are there policies and processes that emphasize efficiency at the expense of diversity?

2. Ecological variability: Resilience is about embracing and working with ecological variability, rather than attempting to control and reduce it. Holding a system in the same (desired) condition erodes resilience because the capacity to absorb disturbance is based on the system's history of dealing with disturbances.

3. Modularity: Resilient systems consist of modular components. In what ways is the system you're interested in modular and is this modularity changing? Is the system becoming more fully connected, or are there parts of it that are becoming more isolated, or too loosely connected?

4. Acknowledging slow variables: There needs to be a focus on the controlling (often slowly changing) variables associated with thresholds. What are the slow variables controlling your system? The "rule of hand" that arose out of comparative studies says that at any one scale there are no more than three to five important controlling variables. It invokes Buzz Holling's call for "requisite simplicity" in attempting to understand and manage social-ecological systems.

5. Tight feedbacks: A resilient world possesses tight feedbacks (but not too tight). Are the signals from cost/benefit feedbacks loosening? Are procedural requirements increasing the time it takes to detect and respond to system changes?

6. Social capital: This is about promoting trust, well-developed social networks, and effective leadership. Sometimes leadership needs to be vested in a strong, visionary individual; at others it needs to be more of a process of shepherding or leading from behind.

7. Innovation: Resilience places an emphasis on learning, experimentation, locally developed rules, and embracing change. Following a disturbance, did the system change its practice in an appropriate manner? Did it learn? Is experimentation being encouraged, or is the government paying subsidies to continue doing the same thing?

8. Overlap in governance: A resilient world would have institutions that include "redundancy" in their governance structures, including a mix of common and private property with overlapping access rights.

9. Ecosystem services: A resilient world includes all the unpriced ecosystem services in development proposals and assessments. In practice that means getting to know your ecosystem services—where they come from, how they are bundled, who benefits and who doesn't, how they might be affected by potential thresholds, how changes in one can influence the resilience of others.

Extending the List

In *Resilience Thinking* we encouraged readers to tell us how they might add to our list of nine attributes. We have received many suggestions. What was fascinating about the responses was that the same words and concepts kept appearing. They fall roughly into four main groups:

- Democratization / distributed governance / decentralization (a bigger emphasis than we gave in our attribute number 8)
- Fairness / equality / "fair trade"
- Humility (a need to accept uncertainty)
- Learning / education about resilience / change of "myths"

In terms of the discussion on global resilience, it seems to us that these four themes might be combined into a tenth and an eleventh attribute of a resilient world:

10. Fairness/equity: A (desirable) resilient world would acknowledge notions of equality among people, would encourage democratization so that everyone has a say, a sense of agency, and would promote the notion and practice of "fair trade." These attributes would encourage diversity, innovation, collaboration, and effective feedbacks while promoting higher levels of social capital.

11. Humility: A resilient world would acknowledge our dependence on the ecosystems that support us, would allow us to appreciate the limits of our mastery and accept that we have much to learn, and would ensure that our people are well educated about resilience and our interconnection with the biosphere.

Even if we adopt these eleven attributes as goals (and achieve them), there's no guarantee that we will sidestep the looming shocks and changes we are facing. However, a resilient world will be better placed, come what may.

Emergent Themes for Effective Practice

We list here a number of themes that have emerged as we wrote this book, based on our own experience and on views shared with us by Resilience Alliance colleagues. As you begin your own practice, it's worth asking whether these themes are apparent, and if you can add to them.

Think Multiple Scales

You cannot understand or manage a system by focusing on one scale. The scale that your system works at is embedded in scales above, up to the global scale. We can't "fix" the global scale without paying attention to necessary changes at scales below, and vice versa; we can't ensure the future well-being of the systems we all care about without paying attention to necessary changes/developments at the global scale.

This puts a special focus on the cross-scale nature of these systems. Our place is local, but it's also part of the region and the globe. We live in all three, resilience is present in all three, but we need to connect all scales.

You may be three or more scales away from the global scale, but your own system is embedded in it; and only by all scales pushing the message up to the global scale will appropriate action at the global scale be taken; part of resilience practice at any scale is to push the message as far up the scale as possible.

Put a Focus on Thresholds

Thresholds frame regimes, and a focus on managing system regimes is useful because it frames the action arena, highlighting two kinds of problems. One is to keep the system in its current regime; the other is to restore it or change it to another regime (if the one you're in is undesirable). Use a variety of approaches to identify known and possible thresholds that you need to take into account. Start by getting to know the ecosystem services—where they come from, how they are bundled.

Celebrate Change

Some people perceive change as good; others fear it. This is one of the creative opposites that are never resolved. We live in the currents set by that interaction. The result is that growth is not continuous but episodic—times of quiet, times of noise and conflict, times of innovations, times of recovery. That is the adaptive cycle writ large.

Organizations require periodic reinvention but inevitably become less flexible and innovative over time. When they do, abandon tired elements, preserve and enhance good elements, and introduce some qualitatively new ones. Past dependencies will resist this, but proceed by making the new steps examples of experimentation, and watch for synergies among them. Those are the sources of innovation.

Embrace Uncertainty

Feedbacks and nonlinearities are generally perceived as the source of surprises. Celebrate the delight in surprise. Surprises come with the unknown; it is the source of unexpected synergies that unlock a shift into a new regime of recovery or of design. As we've previously pointed out, embracing uncertainty keeps us on our toes and acts as a positive influence on being able to cope with the surprises that come. It builds general resilience.

Box 6: Emerging Frontiers

What are the big themes a resilient world needs to be working on, areas that are not getting the emphasis they deserve? Here we nominate three that we believe deserve much more attention than they are currently receiving.

1. Urbanizing regions: Our focus for much of this book, based on current resilience activities, has been on social-ecological systems at local and regional scales—towns, agriculture, and natural ecosystems; villages and fisheries; lakes and swamps. But the future big problems, now emerging all over the developing world, will be in resilience of urbanizing regions—peri-urban regions going through rapid and massive (and often largely unplanned) expansion. They are where more than half of all the people in the world will be living in coming years. Unlike existing, older cities, their development trajectories are not yet fixed—but they are becoming progressively locked in.

We have not addressed urban resilience in this book; it deserves a book of its own. But we cannot finish without mentioning it. These "systems" are social-technical-ecological systems, and putting resilience thinking into practice in these very dynamic systems, learning how to guide their trajectories so as to avoid crossing undesirable thresholds (lock-in traps), and learning about the attributes that confer general resilience on an urban system should be high on the list for those interested in the resilience of the world.

2. New technologies: The individual and interactive consequences of the rapidly developing fields of nano-, bio-, and info-technology for resilience have barely been considered. We can see how they might have both positive and negative effects, but the greatest concern is that perceived immediate benefits will be pursued without consideration of their systemwide (and therefore resilience) consequences. Their greatest threat is that unintended secondary consequences may be impossible to contain or reverse; their greatest promise lies in their potential to help achieve the transformational changes that are required at all scales.

3. General resilience: Much of this book has been about specified resilience—it's what we can measure and get our teeth into. It is very important and its loss is a real concern, especially as reflected in the planetary boundaries. But the biggest challenge and the weakest part of current resilience practice is the erosion of general resilience, worldwide, at all scales. It is our lack of understanding and ability to deal with general resilience that most needs research. One key component of general resilience is diversity, and we are losing that at an alarming rate, in our biodiversity, our cultures, and our ways of doing things.

Foster Innovation

This includes maintaining diversity, free exchange of ideas to promote bricolage, tolerance of oddball ideas along with ability to abandon ideas that do not work out, openings for serendipity, opportunity to experiment and learn from mistakes, capacity to scaffold good ideas rapidly into new platforms, active creation of disturbance, and finding windows of opportunity for grafting subversive thinking onto prevailing institutional structures. There is a need to share experiences in implementing resilience, broaden the learning potential from the experiments of others, and expand these learning networks globally.

And Don't Forget Governance

Navigating the combined influences of exogenous shocks and endogenous changes calls for adaptive governance. Changing governance is often the hardest thing to achieve, and usually the most important.

Resilience and Surprise

A comment we hear quite often is along the lines of "not everything is about resilience." Of course this is true. Improving and sustaining human well-being calls for contributions from many fields. We contend, however, that resilience is a first-order concern, and increasing human well-being within the limits of resilience is a second-order concern. Think of it as seeking a preferred state of the system (second-order requirement) within the preferred stability domain of the system (first-order requirement).

If you imagine all the problems and processes we have to deal with as existing in a space defined by their controllability and their uncertainty, then optimal control approaches are appropriate in the area of high certainty and high controllability. As uncertainty increases and controllability decreases, the likelihood of surprises increases, and so a resilience approach, including adaptive governance and management, assumes greater importance. And there can be little doubt that the world is moving into a space of greater uncertainty and growing surprise.

With a swelling population, increasing movement and connectedness, and climate change and other impacts of technology, we are witnessing an increasing frequency of big surprises—floods, twisters, droughts, and failures of what once were considered fail-safe infrastructure (like the Fukushima nuclear reactors). While some politicians and business leaders deny it, those who are directly affected (the reinsurance industry, for example) do not. Munich Re, a global reinsurance company, reported that 2011 was the highest loss year on record, and that statement was made only halfway through the year!

Setting up resilience as something different from other ways of contributing to improved human well-being unfortunately obscures what it is really all about. Putting resilience into practice begins with putting a resilience lens over existing plans and policies. A huge amount of excellent work involving many disciplines is being done in efforts to improve human well-being. Putting a resilience lens over this comes down to understanding how these efforts interact within a multiscale, complex adaptive system with some likely critical transition points— whatever field of endeavor you are in, whatever changes/developments are being considered. Rather than *in contrast to* or *instead of*, a resilience framework is *complementary to* other ways of approaching the challenge of improving human well-being.

So, while resilience thinking is not a panacea, the more we can effectively put it into practice, the better shape our systems—be they local farms or the whole world—will be in to deal with the looming surprises that lie ahead.

Postscript:
A View from the Northwest Passage

In July 2010, B. W. sailed for six days through the heart of the famous Northwest Passage on a Canadian icebreaker, the *Louis Saint Laurent*. It was part of Canada's Three Oceans research program. A group of scientists, businessmen, and senior government officials were on board for this leg of a summer sampling trip to discuss the changes taking place in the Arctic.

For the first three days we crunched through ice in Barrow Strait and into Peel Sound, noting the patchiness of first-year and multiyear ice. During the third night we broke clear of the ice into open water. It was somewhere near here that Sir John Franklin's third and final expedition disappeared in 1845, still trying to find a way through the ice. On a desolate shore of King William Island sits "a long forgotten lonely cairn of stones" (a line from the famous Stan Rogers song "Northwest Passage"). I was lucky enough to visit that cairn on a short helicopter excursion. We went to replace a plaque that a polar bear had torn off. The plaque commemorates the expedition that came to find Franklin, and then also perished after being trapped in thick ice for nearly two years.

The Northwest Passage is now basically ice-free in summer. The first boats sailed through it nonstop in 2007. Based on current trends it will soon be completely ice-free—several decades before predictions based on climate models.

The Arctic is the harbinger of global change for the world. It's happening there faster than anywhere else, and it provides a valuable insight into what's happening to our planet. It therefore makes an appropriate finale to this book, as it links some of the changes in resilience being witnessed in the Arctic to the lessons we've been encountering for resilience practice. Some of the impacts in the following discussion we witnessed as we traveled through the region; some we learned from experts traveling with us.

The drivers of change in the Arctic are atmospheric warming and rising levels of carbon dioxide. Their interactive effects are leading to a succession of impacts at increasing scales, illustrating the kinds of feedbacks that a resilience assessment would identify.

First and foremost, the warming atmosphere is causing loss of sea ice. This has a positive feedback causing further warming. Ice covered by snow has a high albedo—it reflects solar radiation—while open water absorbs most of the incoming energy. This is the first important feedback change leading to a threshold and regime shift in the Arctic Ocean (from ice cover, to open water), and the threshold was passed in 1996/97. Short of geoengineering a layer of reflective particles high in the atmosphere, there is nothing we can do about it. Whether geoengineering is even possible (or desirable), the threshold would still be there, but the incoming energy would be reduced and therefore kept away from the Arctic.

Our first encounter with the secondary effects of the shift to an ice-free Arctic came from a discussion with the mayor of the small town of Resolute Bay. The fishers and hunters, he told us, had become very wary of venturing across to nearby islands where they traditionally hunted. This was because they were encountering waves, and their small boats, which are easy to manipulate over icy patches, were not designed for waves. Waves are a new phenomenon in the Arctic. In addition to overwhelming small boats, they are causing significant shoreline erosion, and this is happening all across the northern shores of Canada and Alaska. The hunters will likely respond by getting bigger boats, which is likely to bring with it further feedback effects as they change their hunting habits.

At the scale of the Arctic basin, the ice-free state is initiating a series of rapid responses by many countries as new trade routes are developed and as gas and oil exploration experiences a massive increase in activity in the Arctic region. The pattern of circumpolar flows of water means that any negative effects (like an oil spill) would reach almost all of the Arctic Ocean and shoreline.

The warming trend, coupled with increasing levels of carbon dioxide, is also leading to another kind of threshold, one to do with ocean acidity. There is a significant increase in freshwater runoff from Canada and Siberia, and the water running off has very low concentrations of carbonate ions (as does the meltwater from melting sea ice). Add to this the increased carbon dioxide being absorbed into the water, and the ocean is becoming more acidic. The Arctic Ocean has the highest rate of acidification in the world, with significant declines in aragonite (the form of calcium carbonate that plankton need to make their tiny shells). It is affecting the solubility of calcium carbonate. In 2008 the surface water tipped from an environment that enabled the formation of shells to one in which shells dissolve.

There has also been an increase in stratification of the Arctic Ocean, caused by the freshening of the upper layers due to runoff of ice melt from the land, constraining the upward flux of nutrients. This further affects the food web. The net effect of greater acidity, lower arago-nite concentrations, and lower nutrient levels has been a detectable decline in the size of microplankton—from nannoplankton to smaller picoplankton. It is feared this might have a cascading effect up the food chain. It's possible we might see the replacement of polar bears and seals as top predators, by a system with jellyfish living on tiny plankton too small for use by the existing lower-food-chain species.

And it's not just polar bears that have to worry about the changes, because what's happening in the Arctic will affect the rest of the world. The warmer water and air mass in the Arctic interact in a coupled manner, and changes in ocean circulation and temperatures are ex-pected to shift trade winds that bring rains to North America. They are also implicated in causing cold, wet winters in the mid to high temperate regions in the Northern Hemisphere. Enhanced storage of heat in ocean areas (now free of sea ice) means that more heat is re-turned to the atmosphere the following autumn, modifying air pres-sure differences, resulting in a warm Arctic–cold continents pattern. Consider the shocking storms across Europe and America in 2009 and 2010. These were caused by north-south airstreams from the polar re-gion that previously remained circulating around the pole. The conse-quences will be changes in the climate of the Northern Hemisphere, and thus of the world.

The story from the Arctic, therefore, is one of ocean warming, loss of sea ice, permafrost thawing and greenhouse gas release, rising sea lev-el and coastal erosion, altered wind fields and storm tracks, increased river discharge, shifting ocean fronts, invasion by nonindigenous species, increased hypoxia, ocean acidification, and potential impacts on the thermohaline circulation (Carmack and McLaughlin 2011). They all interact and give rise to a number of thresholds—some crossed, some looming—and a confusing picture of societal response.

A year after the Northwest Passage voyage, Sweden, as current chair of the Arctic Council, initiated a project to develop an Arctic Resilience Report, and the initial workshop highlighted three significant things. First, a number of looming social thresholds were identified, as climate and economic changes disrupt traditional lifestyles. At the very time when the general resilience of the people who live in the Arctic region is increasingly important, it is being progressively eroded. Second, responses by the international community are mostly dominated by national self-interest. And third, some parts of the Arctic, and some of its inhabitants, will have to undergo transformational change. It will be done to them, or they can do it deliberately in ways that will, hopefully, allow them to move into a desired future; and in this regard there are some positive possibilities.

But time is running out to make the choices. Activists with vested interests (the denialists) quibble with the science to delay action because they have most to lose in the short term. But what is happening in the Arctic, coupled with the risks (and costs) of increasing global catastrophes, points to the intersection of rising stresses and disturbance, and declining resilience. It calls for a different way of dealing with uncertainty—it puts a high priority on getting resilience thinking into practice.

References

Almedom, A., Tesfamichael, B., Mohammed, Z. S., Mascie-Taylor, C. G., and Alemu, Z. 2007. Use of "sense of coherence (soc)" scale to measure resilience in Eritrea: Interrogating both the data and the scale. *Journal of Biosocial Science* 39:91–107.

Anderies, J. M., Janssen, M. A., and Walker, B. H. 2002. Grazing management, resilience and the dynamics of a fire driven rangeland. *Ecosystems* 5:23–44.

Arrow, K. and Hahn, F. 1971. *General competitive analysis*. Holden-Day, San Francisco.

Ash, A. J., McIvor, J. G., Mott, J. J., and Andrews, M. H. 1997. Building grass castles: Integrating ecology and management of Australia's tropical tallgrass rangelands. *Rangeland Journal* 19:123–144.

Basurto, X. 2008. Biological and ecological mechanisms supporting marine self-governance: The Seri callo de hacha fishery in Mexico. *Ecology and Society* 13. Online at www.ecologyandsociety.org/vol13/iss2/art20/.

Basurto, X. and Coleman, E. 2010. Institutional and ecological interplay for successful self-governance of community-based fisheries. *Ecological Economics* 69:1094–1103.

Bestelmeyer, B. T., Tugel, A. J., Peacock, G. L., Robinett, D. G., Shaver, P. L., Brown, J. R., Herrick, J. E., Sanchez, H., and Havstad, K. M. 2009. State-and-transition models for heterogeneous landscapes: A strategy for development and application. *Rangeland Ecology and Management* 62:1–15.

Biggs, H. C. and Rogers, K. H. 2003. An adaptive system to link science, monitoring, and management in practice. In du Toit, J. T., Rogers, K. H., and Biggs, H. C., eds., *The Kruger experience: Ecology and management of savanna heterogeneity*, 59–80. Island Press, Washington, D.C.

Bodin, Ö., Tengö, M., Norman, A., Lundberg, J., and Elmqvist, T. 2006. The value of small size: Loss of forest patches and threshold effects on ecosystem services in southern Madagascar. *Ecological Applications* 16:440–451.

Bousquet, F., Bakam, I., Proton, H., and Le Page, C. 1998. Cormas: Common-pool resources and multiagent systems. *Lecture Notes in Artificial Intelligence* 1416:826–837.

Bousquet, F., Barreteau, O., Le Page, C., Mullon, C., and Weber, J. 1999. An environmental modelling approach: The use of multi-agents simulations. In Blasco, F. and Weill, A., eds., *Advances in environmental and ecological modelling*, 113–122. Elsevier, Paris.

Briske, D. D., Bestelmeyer, B. T., Stringham, T. K., and Shaver, P. L. 2008. Recommendations for development of resilience-based state-and-transition models. *Rangeland Ecology and Management* 61:359–367.

Brown, K. and Westaway, E. 2011. Agency, capacity, and resilience to environmental change: Lessons from human development, well-being, and disasters. *Annual Review of Environment and Resources* 36:14.1–14.22.

Brown, V. A. 2010. Collective inquiry and its wicked problems. In Brown, V. A., Harris, J. A., and Russell, J. Y. *Tackling wicked problems through the transdisciplinary imagination*, chap. 4. Earthscan, London.

Burke, L., Reytar, K., Spalding, M., and Perry, A. 2011. *Reefs at risk revisited*. World Resources Institute, Washington, D.C.

Campbell, J., Friedman, B. J., Solow, R. M., and Temin, P. 2011. The financial crisis and economic policy. *Bulletin of the American Academy of Arts and Sciences* 64:36–44.

Carmack, E. C. and McLaughlin, F. A. 2011. Towards recognition of physical and geochemical change in subarctic and arctic seas. *Progress in Oceanography* 90:90–104.

Carpenter, S. R. 2003. *Regimes shifts in lake systems*. Ecology Institute, Oldendorf/Luhe, Germany.

Carpenter, S. R. and Bennett, E. M. 2011. Reconsideration of the planetary boundary for phosphorus. *Environmental Research Letters* 6:014009.

Carpenter, S. R. and Brock, W. A. 2006. Rising variance: A leading indicator of ecological transition. *Ecology Letters* 9:311–318.

Carpenter, S. R., Ludwig, D., and Brock, W. A. 1999. Management of eutrophication for lakes subject to potentially irreversible change. *Ecological Applications* 9:751–771.

Carson, J. and Doyle, J. 2000. High optimized tolerance: Robustness and design in complex systems. *Physical Review Letters* 84:2529–2532.

Castilla, J. C., et al. 1998. Artisanal caletas: As units of production and co-managers of benthic invertebrates in Chile. *Canadian Journal of Fishery and Aquatic Science* 125:407–413.

Coe, M. T. and Foley, J. A. 2001. Human and natural impacts on the water resources of the Lake Chad basin. *Journal of Geophysical Research* 106:3349–3356.

Cox, M. and Ross, J. M. 2011. Robustness and vulnerability of community irrigation systems: The case of the Taos Valley acequias. *Journal of Environmental Economics and Management*. doi:10.1016/j.jeem.2010.10.004.

Cumming, D. H. M. 2005. Wildlife, livestock and food security in the South-East Lowveld of Zimbabwe. In *Proceedings of the Southern and East African Experts Panel on Designing Successful Conservation and Development Interventions at the Wildlife/Livestock Interface*, 41–45. IUCN Occasional Paper. IUCN, Gland, Switzerland.

Dakos, V., Scheffer, M., van Nes, E. H., Brovkin, V., Petoukhov, V., et al. 2008. Slowing down as an early warning signal for abrupt climate change. *Proceedings of the National Academy of Sciences* 105:14308–14312.

Dakos, V., van Nes, E. H., Donangelo, R., Fort, H., and Scheffer, M. 2010. Spatial correlation as leading indicator of catastrophic shifts. *Theoretical Ecology* 3:163–174.

Dietz, T., Ostrom, E., and Stern, P. C. 2003. The struggle to govern the commons. *Science* 302:1902–1912.

Easterlin, R. A. 2003. Explaining happiness. *Proceedings of the National Academy of Sciences* 100:11176–11183.

Ellery, W. N., McCarthy, T. S., and Smith, N. D. 2003. Vegetation, hydrology and sedimentation patterns on the major distributary system of the Okavango fan, Botswana. *Wetlands* 23:357–375.

Fernandez, R. J., Archer, E. R. M., Ash, A. J., Dowlatabadi, H., Hiernaux, P. H. Y., Reynolds, J. F., Vogel, C. H., Walker, B. H., and Legand, T. W. 2002. Degradation and recovery in socio-ecological systems: A view from the household/farm level. In Reynolds, J. F. and Stafford-Smith, M., eds., *Global desertification: Do humans cause deserts?* Dahlem University Press, Berlin.

Foley, J. A., DeFries, R., Asner, G. P., et al. 2005. Global consequences of land use. *Science* 309:570–574.

Folke, C., Carpenter, S. R., Walker, B., Scheffer, M., Chapin, T., and Rockström, J. 2010. Resilience thinking: Integrating resilience, adaptability and transformability. *Ecology and Society.* Online at www.ecologyandsociety.org/vol15/iss4/art20/.

Folke, C., Hahn, T., Ollsen, P., and Norberg, J. 2005. Adaptive governance of social-ecological systems. *Annual Review of Environment and Resources* 30:441–473.

Frank, R. 1999. *Luxury fever.* Princeton University Press, Princeton, New Jersey.

Gelcich, S., Hughes, T. P., Olsson, P., Folke, C., Defeo, O., Fernández, M., and Foaleb, S. 2010. Navigating transformations in governance of Chilean marine coastal resources. *Proceedings of the National Academy of Sciences* 107:16794–16799.

Gillson, L. and Duffin, K. I. 2006. Thresholds of potential concern as benchmarks in the management of African savannahs. *Philosophical Transactions of the Royal Society B* 362:309–319.

Granovetter, M. 1978. Threshold models of collective behaviour. *American Journal of Sociology* 83:1420–1443.

Gunderson, L. H., and Holling, C. S., eds. 2002. *Panarchy: Understanding transformations in human and natural systems.* Island Press, Washington, D.C.

Gunderson, L. and Light, S. 2006. Adaptive management and adaptive governance in the Everglades ecosystem. *Policy Sciences* 39:323–334.

Hansen, L. and Sargent, T. 2008. *Robustness*. Princeton University Press, Princeton, New Jersey.

Hardin, G. 1968. The tradegy of the commons. *Science* 162(3859):1243–1248.

Holling, C. S., ed. 1978. *Adaptive environmental assessment and management*. Wiley, London.

Homer-Dixon, T., Biggs, R., Lambin, E., Deutsch, L., Folke, C., Naevdal, E., Peterson, G., Rockström, J., Scheffer, M., Steffen, W., Troell, M., and Walker, B. In prep. The architecture of global crises: Multiple whammies, long fuses, and ramifying cascades.

Hommes, C. H. and Wagener, F. O. O. 2009. Complex evolutionary systems in behavioural finance. In Hens, T. and Reiner Schenk-Hoppe, K., eds., *Handbook of financial markets: Dynamics and evolution*. North-Holland, Amsterdam.

Janssen, M., Walker, B., Langridge, J., and Abel, N. 2000. An adaptive agent model for analysing co-evolution of management and policies in a complex rangeland system. *Ecological Modelling* 131:249–268.

Kahneman, D. and Deaton, A. 2010. High income improves evaluation of life but not emotional well-being. *Proceedings of the National Academy of Sciences* 107:16489–16493.

Keskinen, M., Käkönen, M., Tola, P., and Varis, O. 2007. The Tonle Sap Lake, Cambodia: Water-related conflicts with abundance of water. *The Economics of Peace and Security Journal* 2:48–58.

Kingsford, R. T., Biggs, H. C., and Pollard, S. R. 2011. Strategic adaptive management in freshwater protected areas and their rivers. *Biological Conservation* 144:1194–1203.

Kinzig, A. P., Ryan, P., Etienne, M., Allison, H., Elmqvist, T., and Walker, B. 2006. Resilience and regime shifts: Assessing cascading effects. *Ecology and Society* 11. Online at www.ecologyandsociety.org/vol11/iss1/art20/.

Kok, K. 2009. The potential of fuzzy cognitive maps for semi-quantitative scenario development, with an example from Brazil. *Global Environmental Change* 19:122–133.

Lansing, J. S. and Fox, K. M. 2011. Niche construction on Bali: The gods of the countryside. *Philosophical Transactions of the Royal Society B* 366:927–934.

Levin, S. A. 1999. *Fragile dominion: Complexity and the commons*. Basic Books, New York.

Lorenzen, R. P. and Lorenzen, S. 2010. Changing realities: Perspectives on Balinese rice cultivation. *Human Ecology*. doi:10.1007/s10745-010-9345-z.

Ludwig, J. A. and Tongway, D. J. 1995. Spatial organisation of landscapes and its function in semi-arid woodlands, Australia. *Landscape Ecology* 10:51–63.

Maler, K., Li, C. Z., and Destouni, G. 2007. Pricing resilience in a dynamic economy-environment system: A capital-theoretical approach. *Beijer Discussion Papers* 208, Royal Swedish Academy of Sciences, Stockholm.

Martin, J., Runge, M. C., Nichols, J. D., Lubow, B. C., and Kendall, W. L. 2009. Structured decision making as a conceptual framework to identify thresholds for conservation and management. *Ecological Applications* 19:1079–1090.

Masten, A. S. 2001. Ordinary magic: Resilience processes in development. *American Psychologist* 56:227–238.

Mathevet, R., Le Page, C., Etienne, M., Lefebvre, G., Poulin, B., Gigot, G., Proreol, S., and Mauchamp, A. 2007. BUTORSTAR: A role playing game for collective awareness of wise reedbed use. *Simulation and Gaming* 38:233–262.

McCarty, N., Poole, K., Romer, T., and Rosenthal, H. 2010. Political fortunes: On finance and its regulation. *Daedalus,* Fall 2010.

McKenzie, L. W. 2002. *Classical general equilibrium theory*. MIT Press, Cambridge, Massachusetts.

Micklin, P. 2007. The Aral Sea disaster. *Annual Review of Earth and Planetary Sciences* 35:47–72.

Millennium Ecosystem Assessment. 2005. *Ecosystems and human well-being: Synthesis*. Island Press, Washington, D.C.

NSW Natural Resources Commission. 2009. Regional forest assessment: River red gums and woodland forests. Online at www.nrc.nsw.gov.au/content/ documents/Red%20gum%20-%20FAR%20-%20Rec%20report.pdf.

NSW Natural Resources Commission. 2011. Framework for assessing and recommending upgraded catchment action plans. Online at www.nrc.nsw. gov.au/content/documents/Framework%20for%20CAPs.pdf.

Ostrom, E. 2008. The challenge of common-pool resources. *Environment: Science and Policy for Sustainable Development* 50:8–20.

Ostrom, E. 2009. A general framework for analyzing sustainability of social-ecological systems. *Science* 325:419–422.

Özesmi, U. and Özesmi, S. 2003. A participatory approach to ecosystem conservation: Fuzzy cognitive maps and stakeholder group analysis in Uluabat Lake, Turkey. *Environmental Management* 31:0518–0531.

Panabokke, C. R., Sakthivadivel, R., and Weerasinghe, A. D. 2002. *Evolution, present status and issues concerning small tank systems in Sri Lanka*. International Water Management Institute, Colombo, Sri Lanka.

Peterson, G. D., Cumming, G. S., and Carpenter, S. R. 2003. Scenario planning: A tool for conservation in an uncertain world. *Conservation Biology* 17:358–366.

Pine, W. E., Martell, S. J., Walters, C. J., and Kitchell, J. F. 2009. Counterintuitive responses of fish populations to management actions: Some common causes and implications for predictions based on ecosystem modeling. *Fisheries* 34:165–180.

Pollnac, R., Christie, P., Cinner, J. E., Dalton, T., Daw, T. M., Forrester, G. E., Graham, N. A., and McClanahan, T. R. 2010. Marine reserves as linked social-ecological systems. *Proceedings of the National Academy of Sciences* 107:18262–18265.

Quiggan, J. 2010. *Zombie economics: How dead ideas still walk amongst us.* Princeton University Press, Princeton, New Jersey.

Radford, J. Q., Bennett, A. F., and Cheers, G. J. 2005. Landscape-level thresholds of habitat cover for woodland-dependent birds. *Biological Conservation* 124:317–337.

Rausdsepp-Hearne, C., Peterson, G. D., and Bennett, E. M. 2010. Ecosystem service bundles for analysing tradeoffs in diverse landscapes. *Proceedings of the National Academy of Sciences* 107:5242–5247.

Resilience Alliance. 2010. *Assessing resilience in social-ecological systems: Workbook for practitioners. Version 2.0.* Online at www.resalliance.org/3871.php.

Rietkerk, M., Dekker, S. C., de Ruiter, P. C., and van de Koppel, J. 2004. Self-organized patchiness and catastrophic shifts in ecosystems. *Science* 305:1926–1929.

Rockström, J., et al. 2009. Planetary boundaries: Exploring the safe operating space for humanity. *Ecology and Society.* Online at www.ecologyandsociety.org/vol14/iss2/art32/. (Short version available at *Nature* 461:472–475.)

Room, P. M. and Thomas, P. A. 1985. Nitrogen and establishment of a beetle for biological control of the floating weed *Salvinia* in Papua New Guinea. *Journal of Applied Ecology* 22:139–156.

Rumpff, L., Duncan, D. H., Vesk, P. A., Keith, D. A., and Wintle, B. A. 2010. State-and-transition modelling for adaptive management of native woodlands. *Biological Conservation.* doi:10.1016/j.biocon.2010.10.026.

Scheffer, M. 2009. *Critical transitions in nature and society.* Princeton University Press, Princeton, New Jersey.

Scheffer, M., Bascompte, J., Brock, W. A., Brovkin, V., Carpenter, S. R., et al. 2009. Early-warning signals for critical transitions. *Nature* 53:53–59.

Schluter, M. and Herrfahrdt-Pahle, E. 2011. Exploring resilience and transformability of a river basin in the face of socioeconomic and ecological crisis: An example from the Amudarya River basin, Central Asia. *Ecology and Society* 16. Online at www.ecologyandsociety.org/vol16/iss1/art32/.

Schultz, L., Folke, C., and Olsson, P. 2007. Enhancing ecosystem management through social-ecological inventories: Lessons from Kristianstads Vattenrike, Sweden. *Environmental Conservation* 34:140–152.

Stafford-Smith, M., Horrocks, L., Harvey, A., and Hamilton, C. 2011. Rethinking adaptation for a 4°C world. *Philosophical Transactions of the Royal Society A* 369(1934):196–261.

Stiglitz, J. E. 2002. *Globalization and its discontents*. Penguin, London.

Suding, K. N. and Hobbs, R. J. 2009. Threshold models in restoration and conservation: A developing framework. *Trends in Ecology & Evolution* 24:271–279.

Thomas, K. W. and Kilmann, R. H. 1974. *Thomas–Kilmann conflict mode instrument*. Xicom, Tuxedo Park, New York.

van den Bergh, J. C. 2011. Environment versus growth: A criticism of "degrowth" and a plea for "a-growth." *Ecological Economics* 70:881–890.

Walker, B. H., Abel, N., Anderies, J. M., and Ryan, R. 2009. Resilience, adaptability, and transformability in the Goulburn-Broken Catchment, Australia. *Ecology and Society* 14. Online at www.ecologyandsociety.org/vol14/iss1/art12/.

Walker, B., Barrett, S., Polasky, S., et al. 2009. Looming global-scale failures and missing institutions. *Science* 325:1345–1346.

Walker, B. and Meyers, J. A. 2004. Thresholds in ecological and social-ecological systems: A developing database. *Ecology and Society*. Online at www.ecologyandsociety.org/vol9/iss2/art3.

Walker, B. and Salt, D. 2006. *Resilience thinking*. Island Press, Washington, D.C.

Walker, B. and Westley, F. 2011. Perspectives on resilience to disasters across sectors and cultures. *Ecology and Society* 16. Online at www.ecologyandsociety.org/vol16/iss2/art4/.

Walters, C. J. 1986. *Adaptive management of renewable resources*. McGraw Hill, New York.

Walters, C. 1997. Challenges in adaptive management of riparian and coastal ecosystems. *Conservation Ecology*. Online at www.consecol.org/vol1/iss2/art1/.

Westoby, M., Walker, B., and Noy-Meir, I. 1989. Opportunistic management for rangelands not at equilibrium. *Journal of Range Management* 42:266–274.

Wissel, C. 1984. A universal law of the characteristic return time near thresholds. *Oecologia* 65:101–107.

Zellmer, S. and Gunderson, L. 2009. *Why resilience may not always be a good thing: Lessons in ecosystem restoration from Glen Canyon and the Everglades*. Lexington Books, Lanham, Maryland.

Glossary

Actors: The people who play a role in or have some influence on a social-ecological system. Sometimes referred to as "agents" in other literature.

Adaptability (adaptive capacity): The capacity of actors in a system (people) to manage resilience. This might be to avoid crossing into an undesirable system regime or to succeed in crossing into a desirable one.

Adaptive cycles: A way of describing the progression of social-ecological systems through various phases of organization and function. Four phases are identified: rapid growth, conservation, release, and reorganization. The manner in which the system behaves is different from one phase to the next, with changes in the strength of the system's internal connections, its flexibility, and its resilience.

> **Rapid growth (r)**: A phase in which resources are readily available and entrepreneurial agents exploit niches and opportunities.
>
> **Conservation (K)**: A phase in which resources become increasingly locked up and the system becomes progressively less flexible and responsive to disturbance.
>
> **Release (omega)**: A phase in which a disturbance causes a chaotic unraveling and release of resources.
>
> **Reorganization (alpha)**: A phase in which new actors (species, groups) and new ideas can take hold. It generally leads into another r phase.
>
> The new r phase may be very similar to the previous r phase or may be fundamentally different. The r to K transition is referred to as the fore loop, and the release and reorganization phases are referred to as the back loop. Though most systems commonly move through this sequence of the phases, there are other possible transitions.

Adaptive governance: Governance that changes in response to new circumstances, problems, or opportunities. It encompasses "distributive" governance (passing decision making down to the level where it is most effectively dealt with) and elements of "polycentric" governance (organization of small-, medium-, and large-scale democratic units such that each may exercise independence to make and enforce rules within a circumscribed scope of authority).

Adaptive management: Treating management as a hypothesis coupled to a management "experiment" to test it. It involves an explicit prediction of the outcome of a management intervention before the intervention is made.

Basin of attraction: All the stable states of the system that tend to change toward the attractor. An attractor is a stable state of a system, an equilibrium state that does not change unless it is disturbed. The basin of attraction is often described using the ball-in-the-basin metaphor (see figure 12).

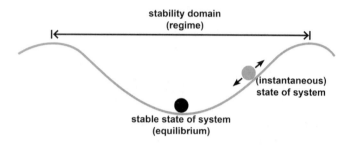

Figure 12: The Ball-in-the-Basin Metaphor

Diversity: The different kinds of components that make up a system. With respect to resilience there are two types of diversity that are particularly important.

Functional diversity: Diversity of the range of functional groups that a system depends on. For an ecological system this might include groups of different kinds of species such as trees, grasses, deer, wolves, and soil. Functional diversity underpins the performance of a system.

Response diversity: Diversity of the range of different response types existing within a functional group. Resilience is enhanced by increased response diversity within a functional group.

Domain: The social, the economic, and the biophysical (ecological) components of a linked system of humans and nature.

Drivers: External forces or conditions that cause a system to change.

Ecosystem services: The combined actions of the species in an ecosystem that perform functions of value to society (e.g., pollination, water purification, flood control).

Equilibrium: A steady-state condition of a dynamic system where the interactions among all the variables (e.g., species) are such that all the forces are in balance and no variables are changing.

Eutrophication: The enrichment of water by nutrients causing an accelerated growth of algae and other plant life.

Feedbacks: The secondary effects of a direct effect of one variable on another that cause a change in the magnitude of that (first) effect. A positive feedback enhances the effect; a negative feedback dampens it.

Governance: The institutions (formal and informal rules, including constitutions, laws, regulations, policies, behavioral rules, and norms) and the organizations, social networks, and social and political processes through which institutions are implemented.

Identity: The essential nature of a system (an individual, an ecosystem, a society) based on the way it functions and on its defining structural characteristics.

Modularity: The degree and pattern of connectedness in a system. A modular system consists of loosely interacting groups of tightly interacting individuals.

Network: The set of connections (number and pattern) between all the actors in a system.

Panarchy: The hierarchical set of adaptive cycles at different scales in a social-ecological system, and their cross-scale effects (i.e., the effects of the state of the system at one scale on the states of the system at other scales). This nesting of adaptive cycles—from small to large—and the influences across scales is referred to as a panarchy.

Regime: A set of states that a system can exist in and still behave in the same way—still have the same identity (basic structure and function). Using the metaphor of the ball in a cup, a regime can be thought of as a system's basin of attraction. Most social-ecological systems have more than one regime in which they can exist.

Regime shift: When a social-ecological system crosses a threshold into an alternate regime of that system.

Resilience: The amount of change a system can undergo (its capacity to absorb disturbance) and remain within the same regime—essentially retaining the same function, structure, and feedbacks.

Robustness: The capacity of a system to perform in a satisfactory way across a defined range of conditions. It overlaps with *resilience* but has a design connotation.

Self-organization: The internal, interactive processes that determine the dynamics of a system, independently of any external influences. A system possessing these processes is a self-organizing system.

Social capital: The capacity of a society to act in a cohesive way to guide its future and to deal with crises. It is based largely on leadership, networks, and its level of trust.

Social-ecological systems: Linked systems of people and nature.

State of a system: Defined by the values of the "state" variables that constitute a system. For example, if a rangeland system is defined by the amounts of grass, shrubs, and livestock, then the state space is the three-dimensional space of all possible combinations of the amounts of these three variables. The dynamics of the system are reflected as its movement through this space.

Sustainability: The likelihood an existing system of resource use will persist indefinitely without a decline in the resource base or in the social welfare it delivers.

System: The set of state variables together with the interactions among them, and the processes and mechanisms that govern these interactions.

Thresholds: Levels in underlying controlling variables of a system at which feedbacks to the rest of the system change.

Transformability: The capacity to create a fundamentally new system (including new state variables, excluding one or more existing state variables, and usually operating at different scales) when ecological, economic, and/or social conditions make the existing system untenable.

Variables (controlling, slow, and fast): Controlling variables in a system (like nutrient levels) determine the levels of other variables that tend to be of concern to people (e.g., algal density and soil fertility). Controlling biophysical variables (e.g., sediment concentration, population age structures) tend to change slowly, while controlling social variables may be fast (e.g., fads) or slow (e.g., culture). In the resilience literature, "slow" variables are often used to mean "controlling" variables. And "fast" variables are often used to imply variables that are of interest to people.

About the Authors

BRIAN WALKER has been one of the leading proponents of resilience theory and practice in the past two decades and currently serves as chair of the board of Resilience Alliance, a multidisciplinary research group that explores the dynamics of complex adaptive systems. Brian is based in Australia at CSIRO Ecosystem Science, and his interests are in the dynamics of linked social-ecological systems. Brian was born and raised in Zimbabwe and obtained his PhD in ecology in Saskatchewan, Canada, in 1968. He lectured at the University of Zimbabwe for six years, was professor of botany and director of the Centre for Resource Ecology at the University of Witwatersrand in Johannesburg for ten years, and then moved to Australia as chief of the CSIRO Division of Wildlife and Ecology, a position he held until 1999.

Brian has played a role in many of the planet's most significant international ecological programs. He led the International Decade of the Tropics Program on Responses of Savannas to Stress and Disturbance from 1984 to 1990 and the Global Change and Terrestrial Ecosystems Project of the International Geosphere-Biosphere Programme from 1989 to 1998. He is a past chair of the board of the Beijer International Institute for Ecological Economics in the Swedish Academy of Science, is a board member of the Stockholm Resilience Centre, and chairs the board of the Australian Research Council Centre of Excellence in Coral Reef Studies. Brian was presented with the Ecological Society of Australia's Gold Medal for 1999. He has coauthored two books, edited six, and written over 200 scientific papers, and he serves on the editorial boards of five international journals.

DAVID SALT has been writing about science, scientists, and the environment for over two decades. Working with CSIRO Education he developed Australia's most popular science magazine for students, called *The Helix*. Following that he served as communication manager for the CSIRO Division of Wildlife and Ecology before becoming the inaugural editor of *Newton* magazine, Australian Geographic's magazine of popular science. He's currently based at the Australian National University in Canberra where he produces *Decision Point*, a national magazine on environmental decision theory. Trained in marine ecology, David is active in environmental education and conservation. He has authored a textbook on farm forestry and biodiversity and coauthored the highly acclaimed *Resilience Thinking* with Brian Walker.

Index

Island Press | Board of Directors

DECKER ANSTROM *(Chair)*
Board of Directors
Comcast Corporation

KATIE DOLAN *(Vice-Chair)*
Conservationist

PAMELA B. MURPHY *(Treasurer)*

CAROLYN PEACHEY *(Secretary)*
President
Campbell, Peachey & Associates

STEPHEN BADGER
Board Member
Mars, Inc.

MARGOT PAUL ERNST
New York, New York

RUSSELL B. FAUCETT
CEO and Chief Investment
Officer, Barrington Wilshire LLC

MERLOYD LUDINGTON LAWRENCE
Merloyd Lawrence, Inc.
   and Perseus Books

WILLIAM H. MEADOWS
President
The Wilderness Society

DRUMMOND PIKE
Founder, Tides
Principal, Equilibrium Capital

ALEXIS G. SANT
Managing Director
Persimmon Tree Capital

CHARLES C. SAVITT
President
Island Press

SUSAN E. SECHLER
President
TransFarm Africa

VICTOR M. SHER, ESQ.
Principal
Sher Leff LLP

SARAH SLUSSER
Executive Vice President
GeoGlobal Energy LLC

DIANA WALL, PH.D.
Director, School of Global
Environmental Sustainability
   and Professor of Biology
Colorado State University

WREN WIRTH
President
Winslow Foundation